McGraw-Hill
My Math

CCSS

# Interactive Guide

## Teacher Edition

Grade 2

Mc
Graw
Hill
Education

# ConnectED.mcgraw-hill.com

**STEM** McGraw-Hill is committed to providing instructional materials in Science, Technology, Engineering, and Mathematics (STEM) that give all students a solid foundation, one that prepares them for college and careers in the 21st century.

Send all inquiries to:
McGraw-Hill Education
8787 Orion Place
Columbus, OH 43240

Selections from:
ISBN: 978-0-02-132716-4 *(Grade 2 Student Edition)*
MHID: 0-02-132716-5 *(Grade 2 Student Edition)*
ISBN: 978-0-02-130122-5 *(Grade 2 Teacher Edition)*
MHID: 0-02-130122-0 *(Grade 2 Teacher Edition)*

Printed in the United States of America.

Visual Kinesthetic Vocabulary® is a registered trademark of Dinah-Might Adventures, LP.

3 4 5 6 7 8 9 ROV 20 19 18 17 16

# Contents

# Chapter 4 Subtract Two-Digit Numbers

# Chapter 5 Place Value to 1,000

# Chapter 6 Add Three-Digit Numbers

# Chapter 7 Subtract Three-Digit Numbers

v

# Chapter 12 Geometric Shapes and Equal Shares

# Proficiency Level Descriptors

| | Interpretive (Input) | | Productive (Output) | |
|---|---|---|---|---|
| | **Listening** | **Reading** | **Writing** | **Speaking** |
| **An Emerging Level EL**<br><br>• New to this country; may have memorized some everyday phrases like, "Where is the bathroom", "My name is….", may also be in the "silent stage" where they listen to the language but are not comfortable speaking aloud<br><br>• Struggles to understand simple conversations<br><br>• Can follow simple classroom directions when overtly demonstrated by the instructor | • Listens actively yet struggles to understand simple conversations<br><br>• Possibly understands "chunks" of language; may not be able to produce language verbally | • Reads familiar patterned text<br><br>• Can transfer Spanish decoding somewhat easily to make basic reading in English seem somewhat fluent; comprehension is weak | • Writes labels and word lists, copies patterned sentences or sentence frames, one- or two-word responses | • Responds non-verbally by pointing, nodding, gesturing, drawing<br><br>• May respond with yes/no, short phrases, or simple memorized sentences<br><br>• Struggles with non-transferable pronunciations. |
| **An Expanding Level EL**<br><br>• Is dependent on prior knowledge, visual cues, topic familiarity, and pretaught math-related vocabulary<br><br>• Solves word problems with significant support<br><br>• May procedurally solve problems with a limited understanding of the math concept. | • Has ability to understand and distinguish simple details and concepts of familiar/ previous learned topics | • Recognizes obvious cognates<br><br>• Pronounces most English words correctly, reading slowly and in short phrases<br><br>• Still relies on visual cues and peer or teacher assistance | • Produces writing that consists of short, simple sentences loosely connected with limited use of cohesive devices<br><br>• Uses undetailed descriptions with difficulty expressing abstract concepts | • Uses simple sentence structure and simple tenses<br><br>• Prefers to speak in present tense. |
| **A Bridging Level EL**<br><br>• May struggle with conditional structure of word problems<br><br>• Participates in social conversations needing very little contextual support<br><br>• Can mentor other ELs in collaborative activities. | • Usually understands longer, more elaborated directions, conversations, and discussions on familiar and some unfamiliar topics<br><br>• May struggle with pronoun usage | • Reads with fluency, and is able to apply basic and higher-order comprehension skills when reading grade-appropriate text | • Is able to engage in writing assignments in content area instruction with scaffolded support<br><br>• Has a grasp of basic verbs, tenses, grammar features, and sentence patterns | • Participates in most academic discussions on familiar topics, with some pauses to restate, repeat, or search for words and phrases to clarify meaning. |

# Strategies for EL Success

Surprisingly, content instruction is one of the most effective methods of acquiring fluency in a second language. When content is the learner's focus, the language used to perform the skill is not consciously considered. The learner is thinking about the situation, or how to solve the problem, not about the grammatical structure of their thoughts. Attempting skills in the target language forces the language into the subconscious mind, where useable language is stored. A dramatic increase in language integration occurs when multiple senses are involved, which causes heightened excitement, and a greater student investment in the situation's outcome. Given this, a few strategies to employ during EL instruction that can make teaching easier and learning more efficient are listed below:

- Activate EL prior knowledge and cultural perspective
- Use manipulatives, realia, and hands-on activities
- Identify cognates
- Build a Word Wall
- Modeled talk
- Choral responses
- Echo reading
- Provide sentence frames for students to use
- Create classroom anchor charts
- Utilize translation tools (i.e. Glossary, eGlossary, online translation tools)
- Anticipate common language problems

## Common Problems for English Learners

| | Cantonese | Haitian Creole | Hmong | Korean | Spanish | Vietnamese |
|---|---|---|---|---|---|---|
| **Phonics Transfers** | | | | | | |
| Pronouncing the /k/as in cake | ● | | ● | ● | | |
| Pronouncing the digraph /sh/ | ● | | ● | | ● | ● |
| Hearing and reproducing the /r/, as in *rope* | ● | | ● | ● | | ● |
| /j/ | | | ● | | ● | |
| Hearing or reproducing the short /u/ | | ● | ● | | | |
| **Grammar Transfers** | | | | | | |
| Adjectives often follow nouns | | ● | ● | | ● | ● |
| Adjectives and adverb forms are interchangeable | | ● | ● | | | |
| Nouns have feminine or masculine gender | | | | | ● | |
| There is no article or there is no difference between articles *the* and *a* | | ● | ● | | | ● |
| Shows comparative and superlative forms with separate words | | | ● | | ● | |
| There are no phrasal verbs | | | | ● | ● | |

# How to Use the Teacher Edition

The Interactive Guide provides scaffolding strategies and tips to strengthen the quality of mathematics instruction. The suggested strategies, activities, and tips provide additional language and concept support to accelerate English learners' acquisition of academic English.

## English Learner Instructional Strategy

Each lesson – including Problem Solving – references an English Learner Instructional Strategy that can be utilized before or during regular class instruction. These strategies specifically support the Teacher Edition and scaffold the lesson for English learners (ELs).

Categories of the scaffolded support are:
- Vocabulary Support
- Language Structure Support
- Sensory Support
- Graphic Support
- Collaborative Support

The goal of the scaffolding strategies is to make each individual lesson more comprehensible for ELs by providing visual, contextual and linguistic support to foster students' understanding of basic communication in an academic context.

## Lesson 1 Make a Hundred to Add
### English Learner Instructional Strategy

**Collaborative Support: Native Language Peers/Mentors**

Review the term *take apart* with students. Explain they will take apart numbers to make a hundred to make it easier to add. Use the online base-ten blocks, in Virtual Manipulatives to assist students visually as they model along with you during the Explore and Explain activity.

Pair bridging level students with same native language emerging/expanding students. Have the pairs work together throughout the lesson. Invite them to explain their thinking with the sentence frames: **I took apart ____ to get ____ + ____. I added ____ and ____ to make a hundred. Then I added ____ + ____ to get the sum, ____.**

Since ELs benefit from visual references to new vocabulary, many of the English Learner Instruction Strategies suggest putting vocabulary words as well as Spanish cognates on a Word Wall. Choose a location in your classroom for your Word Wall, and organize the words by chapter, by topic, by Common Core domain, or alphabetically.

# How to Use the Teacher Edition *continued*

## English Language Development Leveled Activities

These activities are tiered for Emerging, Expanding, and Bridging leveled ELs. Activity suggestions are specific to the content of the lesson. Some activities include instruction to support students with lesson-specific vocabulary they will need to understand the math content in English, while other activities teach the concept or skill using scaffolded approaches specific to ELs. The activities are intended for small group instruction, and can be directed by the instructor, an aide, or a peer mentor.

**English Language Development Leveled Activities**

| Emerging Level | Expanding Level | Bridging Level |
|---|---|---|
| **Build Background Knowledge** Display a number line to 10. ... ... each number ... and then leave your finger on the four. Say, *Now I will count on from four.* Count on three more from four. Say, *I counted on three more from four. What number is it?* Have students answer chorally or with a gesture. **seven** Repeat the activity with another addition problem. | **Public Speaking Norms** Display a number line. Say, *To find the sum of 2 + 5, I started with the greater addend five, and then I counted on two more. The answer is seven.* Emphasize /d/ at the end ... tense words. Ha... volunteer solve ... problem, and then ...se the sentence frame: **I started with _____, and then I counted on _____ more. The answer is _____.** Ensure students emphasize /d/ for past tense. | **Cooperative Learning** Divide students into groups of three. Give each group a number line and 12–15 color tiles or buttons. Say, *Write an addition sentence using the tiles. Use your number ... on and find ...* Give students ...tes to complete the task. Then have one student in each group share the problem they created and explain how they used *counting on* to find the answer. |

*Teacher talk is gray italicized.*

**Student talk is boldfaced.**

**Multicultural Teacher Tip**
Individual praise and standing out from the crowd are frowned upon in some cultures. Students from these cultures may act embarrassed or uncomfortable when invited to the board to solve a problem or asked to share an answer with the class. They may be more comfortable during group classroom activities, and may prefer others in the group to share answers or demonstrate problem solving.

## Multicultural Teacher Tip

These tips provide insight on academic and cultural differences you may encounter in your classroom. While math is the universal language, some ELs may have been shown different methods to find the answer based on their native country, while cultural customs may influence learning styles and behavior in the classroom.

# What's the Math in this Chapter?

## Mathematical Practice

The goals of the Mathematical Practice activity are to help students clarify the specific language of the Mathematical Practice and rewrite the Mathematical Practice in simplified language that students can **relate** to. Examples are discussed to make connections, showing students how they use this Mathematical Practice to solve math problems.

### Chapter 6 Add Three-Digit Numbers

*What's the Math in This Chapter?*

**Mathematical Practice 8: Look for and express regularity in repeated reasoning**

Distribute base-ten blocks to pairs of students and write the addition problem 316 + 472 vertically on the board. Have each pair member model one of the numbers using their base-ten blocks. Once each pair has a model representing 316 and a model representing 472, ask students to combine their like pieces. Say, *How many ones, tens, and hundreds are in each pile?* **7 hundreds, 8 tens, and 8 ones.** Record on the board.

Discuss with students how this is similar to adding two-digit numbers together. They added like place-value together. Model solving the problem on the board in a place-value chart. Highlight the connection between the answer that they found with their base-ten blocks to the traditional algorithm. Remind students [they can use this method] with any addition problem, [beginning with the ones place and regrouping] when needed.

Display a chart with Mathematical Practice 8 and have students assist in rewriting it as an "I can" statement, for example: **I can notice when I need to repeat calculations to solve a problem.** Post the new "I can" statement.

> Mathematical Practice is rewritten as an "I can" statement.

*Inquiry of the Essential Question:*

**How can I add three-digit numbers?**

Inquiry Activity Target: **Students come to a conclusion that they can use repeated calculations to add larger numbers.**

[At the beginning of] the chapter, present the Essential Question to [students. The inquiry] graphic organizer will offer opportunities for students [to observe, make infer]ences, and apply prior knowledge of the addition [algorithm repre]senting the Essential Question. As they investigate, encourage students to draw, write, and collaborate with peers to demonstrate their observations and thinking. Then have students present additional questions they may have to a peer to extend discussions.

> Inquiry Activity Target connects Mathematical Practice to Essential Question.

## Inquiry of the Essential Question

As an introduction to the Chapter, the Inquiry of the Essential Question graphic organizer activity is designed to introduce the Essential Question. The activity offers opportunities for students to observe, make inferences, and apply prior knowledge of samples/models representing the Essential Question. Collaborative conversations drive students toward the Inquiry Activity Target which is to make a connection between the "Mathematical Practice of the chapter" and the "Essential Question of the chapter."

# How to Use the Student Edition

Each student page provides EL support for vocabulary, note taking, and writing skills. These pages can be used before, during, or after each classroom lesson. A corresponding page with answers is found in the Teacher Edition.

## Inquiry of the Essential Question

Students observe, make inferences, and apply prior knowledge of chapter specific samples/models representing the Essential Question of the chapter. Encourage students to have collaborative conversations as they share their ideas and questions with peers. As students inquire the math models, present specific questions that will drive students toward the Inquiry Activity Target which is stated on the Teacher Edition page.

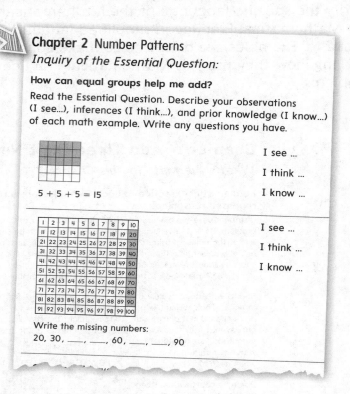

**Chapter 2** Number Patterns
*Inquiry of the Essential Question:*

**How can equal groups help me add?**
Read the Essential Question. Describe your observations (I see...), inferences (I think...), and prior knowledge (I know...) of each math example. Write any questions you have.

5 + 5 + 5 = 15

I see ...
I think ...
I know ...

| 1 | 2 | 3 | 4 | 5 | 6 | 7 | 8 | 9 | 10 |
| 11 | 12 | 13 | 14 | 15 | 16 | 17 | 18 | 19 | 20 |
| 21 | 22 | 23 | 24 | 25 | 26 | 27 | 28 | 29 | 30 |
| 31 | 32 | 33 | 34 | 35 | 36 | 37 | 38 | 39 | 40 |
| 41 | 42 | 43 | 44 | 45 | 46 | 47 | 48 | 49 | 50 |
| 51 | 52 | 53 | 54 | 55 | 56 | 57 | 58 | 59 | 60 |
| 61 | 62 | 63 | 64 | 65 | 66 | 67 | 68 | 69 | 70 |
| 71 | 72 | 73 | 74 | 75 | 76 | 77 | 78 | 79 | 80 |
| 81 | 82 | 83 | 84 | 85 | 86 | 87 | 88 | 89 | 90 |
| 91 | 92 | 93 | 94 | 95 | 96 | 97 | 98 | 99 | 100 |

I see ...
I think ...
I know ...

Write the missing numbers:
20, 30, ___, ___, 60, ___, ___, 90

**Lesson 7** Note Taking
*Compare Numbers to 1,000*

Read the question. Write words you need help with. Use your lesson to write your Cornell notes. Write or draw math examples to explain your thinking.

| Building on the Essential Question | Notes: |
|---|---|
| How can I use a place-value chart to compare numbers to 1,000? | Write the numbers being ordered on a place-value ___.

| hundreds | tens | ones |
|---|---|---|
| 7 | 3 | 2 |
| 7 | 3 | 7 |

Compare the digits in the greatest ___

If the digits are the same, ___ the digits in the next place value.

Continue to compare until the digits are ___.

Then compare using ___ (>), ___ ___ (<), or equal to (=).

732 ◯ 737 |
| Words I need help with: | |

My Math Examples:

## Cornell Notes/Note Taking

Cornell notes offer students a method to use to take notes, thereby helping them with language structure. Scaffolded sentence frames are provided for students to fill in important math vocabulary by identifying the correct word or phrase according to context. Encourage students to refer to their books to locate the words needed to complete the sentences. Each note taking graphic organizer will support students in answering the Building on the Essential Question.

## Four-Square Vocabulary

Four-square vocabulary reinforces the lesson by having students write a definition utilizing the Glossary, write a sentence using the vocabulary in context, and create an example of the vocabulary. Suggest that students use translation tools and write notes in English or in their native language on the graphic organizer as well for clarification of terms. A blank Four-Square Vocabulary template has been provided on p. xix for use with other lessons.

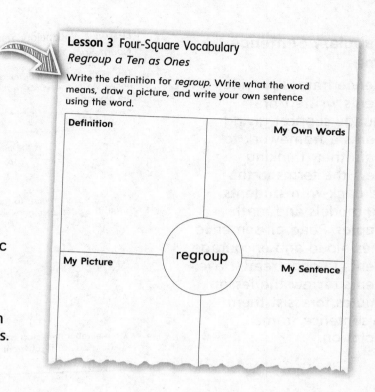

**Lesson 3** Four-Square Vocabulary
*Regroup a Ten as Ones*

Write the definition for *regroup*. Write what the word means, draw a picture, and write your own sentence using the word.

| Definition | My Own Words |
|---|---|
| My Picture | My Sentence |

regroup

**Lesson 4** Guided Writing
*Subtract From a Two-Digit Number*

**How do you subtract from a two-digit number?**
Use the exercises below to help you build on answering the Essential Question. Write the correct word or phrase on the lines provided.

1. What key words do you see in the question?

| Word Bank | | | |
|---|---|---|---|
| ones | regroup | tens | two-digit number |

2. A digit is a symbol used to write numbers. The number 73 is a _____-_____ .

Find 73 − 28.

| | tens | ones |
|---|---|---|
| | 7 | 3 |
| − | 2 | 8 |

3. Can you subtract the ones? _____ You must _____ I ten as 10 ones.

4. Next you should subtract the _____. Then subtract the _____.

5. How do you subtract from a two-digit number?

_____

_____

## Guided Writing

Guided writing provides support to help ELs meet the stated Lesson Objective. Content specific questions are scaffolded to build language knowledge in order to answer the question. Give bridging level students the opportunity to mentor and assist emerging and expanding level students when answering the questions, if needed.

# How to Use the Student Edition *continued*

## Vocabulary Sentence Frame

Sentence frames provide students written/oral language support giving students a framework to explain their thinking. Review the terms in the word back with students using models and math examples. Read all sentence frames aloud and encourage students to echo read. Have students review the lesson examples to assist them with sentence frame completion.

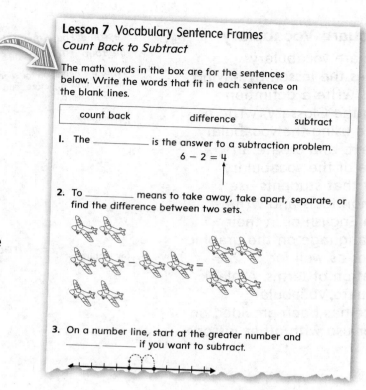

**Lesson 7** Vocabulary Sentence Frames
*Count Back to Subtract*

The math words in the box are for the sentences below. Write the words that fit in each sentence on the blank lines.

| count back | difference | subtract |

1. The _____ is the answer to a subtraction problem.

   $6 - 2 = 4$

2. To _____ means to take away, take apart, separate, or find the difference between two sets.

3. On a number line, start at the greater number and _____ _____ if you want to subtract.

## Concept Web

Concept webs are designed to show relationships between concepts and to make connections. As each concept web is unique in design, please read and clarify directions for students. Encourage students to look through the lesson pages to find examples or words they can use to complete the web.

**Lesson 3** Concept Web
*Add to a Two-Digit Number*

Use the concept web to show what you know about regrouping. Circle *yes* or *no*.

| tens | ones |

25
+ 6

yes   no

yes   no          yes   no

Do I need to regroup
ones as a ten?

32 + 7          | tens | ones |          47
+ 9

## Definition Map

The definition maps are designed to address a single vocabulary word, phrase, or concept. Students should use the Glossary to help define the word in the description box. Sentence frames are provided to scaffold characteristics from the lesson. Students can refer to the lesson examples and Glossary to assist them in completing the sentence frames as well as creating their own math examples. Make sure you review with students the tasks required. A blank Vocabulary Definition Map template can be found on page xvii for use with other lessons.

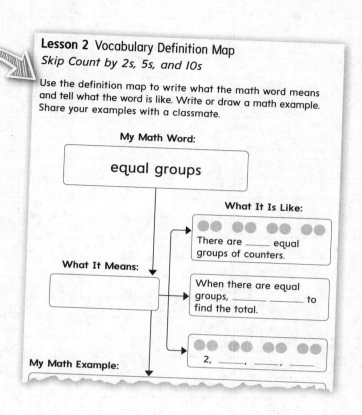

**Lesson 2** Vocabulary Definition Map

*Skip Count by 2s, 5s, and 10s*

Use the definition map to write what the math word means and tell what the word is like. Write or draw a math example. Share your examples with a classmate.

**My Math Word:**

equal groups

**What It Is Like:**

There are ____ equal groups of counters.

**What It Means:**

When there are equal groups, ____ to find the total.

2, ____, ____, ____

**My Math Example:**

## Problem Solving

Each Problem Solving page focuses on scaffolding Exercises 1 and 2 from the Apply the Strategy portion of the lesson in the book. The text for each exercise highlights signal words and phrases to help students decipher the key information in the problem. Visual images as well as tables and part-part-whole mats are included to assist students with the problem-solving process. Sentence frames of the restated question are provided for student oral practice. A blank Problem Solving template can be found on page xx in this book if students need additional assistance with other exercises.

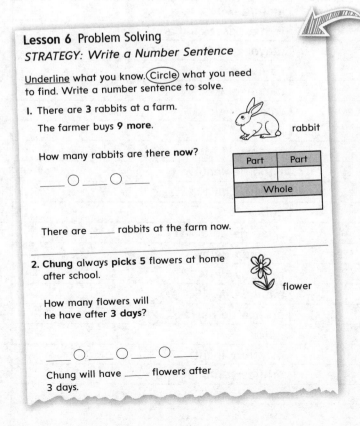

**Lesson 6** Problem Solving

*STRATEGY: Write a Number Sentence*

<u>Underline</u> what you know. (Circle) what you need to find. Write a number sentence to solve.

1. There are **3** rabbits at a farm.

   The farmer buys **9 more**.

   How many rabbits are there **now**?

   ____ ◯ ____ ◯ ____

   rabbit

   | Part | Part |
   |------|------|
   | Whole ||

   There are ____ rabbits at the farm now.

2. **Chung** always **picks 5** flowers at home after school.

   How many flowers will he have after **3 days**?

   flower

   ____ ◯ ____ ◯ ____ ◯ ____

   Chung will have ____ flowers after 3 days.

# English/Spanish Used in Grade 2

| Chapter | English | Spanish |
|---|---|---|
| 1 | add | sumar |
| | addend | sumando |
| | count back | contar hacia atrás |
| | count on | seguir contando |
| | difference | diferencia |
| | doubles | dobles |
| | fact family | familia de operaciones |
| | missing addend | sumando desconocido |
| | near doubles | casi dobles |
| | related facts | operaciones relacionadas |
| | subtract | restar |
| | sum | suma |
| 2 | array | arreglo |
| | equal groups | grupos iguales |
| | even number | número par |
| | odd number | números impares |
| | repeated addition | suma repetida |
| | skip count | contar salteado |
| 3 | regroup | reagrupar |
| 5 | compare | comparar |
| | digit | dígito |
| | equal to (=) | igual a (=) |
| | expanded form | forma desarrollada |
| | greater than (>) | mayor que (>) |
| | hundreds | centenas |
| | less than (<) | menor que (<) |
| | place value | valor posicional |
| | thousands | millares |
| 8 | cent | centavo |
| | cent sign (¢) | signo de centavo (¢) |
| | dime | moneda de 10¢ |
| | dollar | dólar |
| | dollar sign ($) | signo de dólar ($) |
| | nickel | moneda de 5¢ |
| | penny | moneda de 1¢ |
| | quarter | moneda de 25¢ |
| 9 | bar graph | gráfica de barras |
| | data | datos |
| | line plot | diagram lineal |
| | picture graph | gráfica con imágenes |
| | survey | encuesta |

| Chapter | English | Spanish |
|---|---|---|
| | symbol | símbolo |
| | tally marks | marca de conteo |
| 10 | A.M. | a.m. |
| | analog clock | reloj analógico |
| | clock | reloj |
| | digital clock | relog digital |
| | half hour | media hora |
| | half past | y media |
| | hour | hora |
| | hour hand | manecilla horaria |
| | minute | minuto |
| | minute hand | minutero |
| | P.M. | p.m. |
| | quarter hour | cuarto de hora |
| 11 | centimeter | centímetro (cm) |
| | estimate | estimar |
| | foot | pie |
| | inch | pulgada |
| | length | longitud |
| | measure | medir |
| | meter | metro (m) |
| | yard | yarda |
| 12 | cone | cono |
| | cube | cubo |
| | cylinder | cilindro |
| | edge | arista |
| | face | cara |
| | fourths | curators |
| | halves | mitades |
| | hexagon | hexágono |
| | parallelogram | paralelogramo |
| | partition | separar |
| | pentagon | pentágono |
| | pyramid | pirámide |
| | rectangular prism | prisma rectangular |
| | sphere | esfera |
| | thirds | tercios |
| | three-dimensional shape | figura tridimensional |
| | trapezoid | trapecio |
| | two-dimensional shape | figura bidimensional |
| | vertex | vértice |

# Vocabulary Definition Map

Use the definition map to write a description and list characteristics about the vocabulary word or phrase. Write or draw a math example. Write or draw a math example.

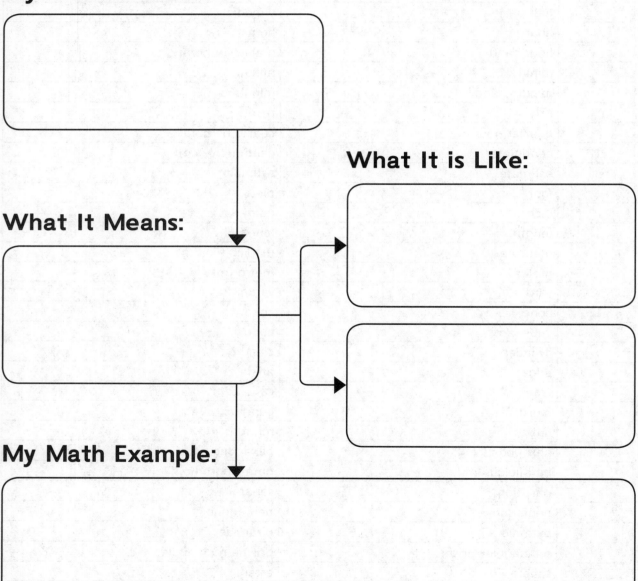

**My Math Word:**

**What It Means:**

**What It is Like:**

**My Math Example:**

**Teacher Directions:** Provide a description, explanation, or example of the new term using images or real objects. Have students use the lesson or Glossary to define the math term. Direct students to list characteristics, and draw a picture representing their math term. Then encourage students to describe their picture to a peer.

# Four-Square Vocabulary

Write the definition for the math word, and what the word means in your own words. Draw a picture that shows the word meaning, then write your own sentence using the word.

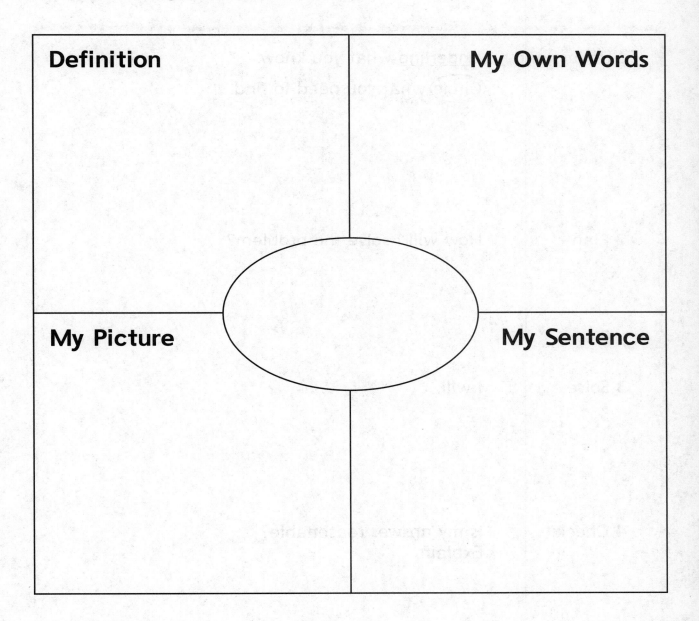

| Definition | My Own Words |
| --- | --- |
| **My Picture** | **My Sentence** |

 **Teacher Directions:** Provide a description, explanation, or example of the new term using images or real objects. Have students use the Glossary to write the definition. Direct students to write a definition in their own words and draw a picture representing their math term. Have students write a sentence using the term and then encourage students to read their sentence to a peer.

# Problem Solving

_____
_____
_____
_____

**1** Understand   <u>Underline</u> what you know.
                 (Circle) what you need to find.

**2** Plan          How will I solve the problem?

**3** Solve         I will...

**4** Check        Is my answer reasonable?
Explain.

# Chapter 1 Apply Addition and Subtraction Concepts

## What's the Math in This Chapter?

**Mathematical Practice 7: Look for and make use of structure**

Show students a group of 4 red crayons and a group of 5 blue crayons. Ask, *How could we find out how many crayons there are all together?* Some students may suggest counting and others may suggest adding. Ask, *What are two addition facts we can use to find the total number of crayons?* **5 + 4 and 4 + 5**. Model that the crayons can be moved easily to visually show the order of both addition facts. Describe this order as the *structure*.

Ask, *Why is the sum the same when you add 5 + 4 and 4 + 5?* **The sum is the same because you are adding the same numbers.** Have students demonstrate understanding by modeling 2 + 3 and 3 + 2 with crayons or other classroom objects. Say, *You can add numbers in any order. The sum is the same. This is the Commutative Property*. The goal of the discussion is for students to notice that when adding, the addends can be added in any order which is a pattern. They are joining two sets of objects together and utilizing the Commutative Property of Addition.

Display a chart with Mathematical Practice 7. Restate Mathematical Practice 7 and have students assist in rewriting it as an "I can" statement, for example: **I can see and use patterns to add in any order.** Have students draw or write examples of the Commutative Property of Addition. Post the examples and new "I can" statement.

## Inquiry of the Essential Question:

### What strategies can I use to add and subtract?

Inquiry Activity Target: **Students come to a conclusion that they can find patterns when they add and subtract.**

As an introduction to the chapter, present the Essential Question to students. The inquiry graphic organizer will offer opportunities for students to observe, make inferences, and apply prior knowledge of patterns and structure representing the Essential Question. As they investigate, encourage students to draw, write, and collaborate with peers to demonstrate their observations and thinking. Then have students present additional questions they may have to a peer to extend discussions.

Regroup students and restate Mathematical Practice 7 and the Essential Question. Pose questions to reflect on what has been learned to guide students in making connections between the Mathematical Practice and the Essential Question.

NAME _____ DATE _____

# Chapter 1 Apply Addition and Subtraction Concepts
## *Inquiry of the Essential Question:*

### What strategies can I use to add and subtract?

Read the Essential Question. Describe your observations
(I see...), inferences (I think...), and prior knowledge (I know...)
of each math example. Write any questions you have.

Fact Family

$9 + 4 = 13$        $13 - 9 = 4$
$4 + 9 = 13$        $13 - 4 = 9$

I see ...

I think ...

I know ...

---

Start at 8. Count back 5.

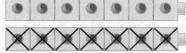

$8 - 5 = 3$

I see ...

I think ...

I know ...

---

Use a doubles fact to find $14 - 7$.

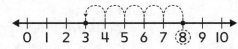

If you know that $7 + 7 = 14$, then $14 - 7 = 7$.

I see ...

I think ...

I know ...

---

Questions I have...

_____

_____

**Teacher Directions:** Read the Essential Question for students. Have students echo read.
Direct students to describe their observations, inferences, and prior knowledge of each
math example. Encourage students to write or draw additional questions they may have.
Then have students share their ideas/questions with a peer.

**Grade 2 · Chapter 1** *Apply Addition and Subtraction Concepts*   1

# Lesson 1 Addition Properties

## English Learner Instructional Strategy

### Vocabulary Support: Frontload Vocabulary

Before the lesson, write *sum* and its Spanish cognate, *suma*, on a cognate chart. Introduce the words *add, addend,* and *sum,* write a math example, and provide concrete objects to support understanding. Have students use an index card to write *sum,* it's cognate, and draw a visual example. Next, have students write a definition and sentence using *sum* in context. Have pairs present a labeled addition number sentence and report back using academic vocabulary. Provide this sentence frame: **If you add the addends, ____ and ____, the sum is____.**

### English Language Development Leveled Activities

| Emerging Level | Expanding Level | Bridging Level |
|---|---|---|
| **Word Knowledge** | **Building Oral Language** | **Public Speaking Norms** |
| Write *add*. Show 3 color tiles and count them. Then show 4 more tiles, and count them. Say, *I will* **add** *3 and 4.* Write 3 + 4 = ____ on the board. Model solving the problem. Then point to each number and symbol as you say, *3 plus 4 is 7.* Switch the tiles' positions making 4 plus 3 and ask, *Will I add to find the answer?* Have students respond chorally or with a gesture, **yes** and then say, *You will add.* Repeat modeling to solve the problem. | Write *sum* and its Spanish cognate, *suma* on a cognate chart. Write some and sum. Display counters. Say, *I will choose some counters.* Point to some. Pick 4 counters, count them, and write 4. Say, *I will choose some more counters.* Repeat with 5 counters. Say, *I will find how many total counters I have.* Write 4 + 5 = ____. Solve the problem. Point to sum, explaining that sum is the answer to an addition problem. Ask, *What is the sum of 4 plus 5?* **The sum of 4 + 5 is 9.** | Write *sum* and *addend* on the board. Have students work in pairs. Give each pair an index card that demonstrates the Commutative Property or the Identity Property of addition. Say, *In your own words, explain the rule shown on your card. Use sum or addend in your explanation.* Give students a few minutes to form a response. Then have one student from each pair share their explanations. For example, **Addends can be added in any order, and the answer will be the same.** |

**Teacher Notes:**

NAME _____ DATE _____

# Lesson I Vocabulary Sentence Frames
## *Addition Properties*

The math words in the box are for the sentences below. Write the words that fit in each sentence on the blank lines.

| add | addend | sum |
|---|---|---|

**I.** Any numbers or quantities being added together are called <u>addends</u>.

$$1 + 3 = 4$$

**2.** The <u>sum</u> is the answer to an addition problem.

$$1 + 3 = 4$$

**3.** To <u>add</u> means to join together sets to find the total.

 **Teacher Directions:** Provide a description, explanation, or example of the each term using images or real objects. Read each sentence frame and have students echo read. Direct students to write the correct terms in each blank. Then encourage students to read each sentence to a peer.

# Lesson 2 Count On to Add
## *English Learner Instructional Strategy*

### Language Structure Support: Multiple-Meaning Word

Before the lesson say, *I am your teacher. I am here to help you. You can count on me.* Share a time when you had to count on someone. Then say, *I counted on _____.* Ask volunteers to share a time when they counted on someone. Provide a sentence frame, such as: **I counted on _____ when _____.** Say, *Count on is used in math to describe a way to add two numbers together.* Display a number line, and write 4 + 2 = _____ on the board. Say, *I can start with four and then count on two more to find the answer.* Demonstrate the process. During the lesson, display purple and green crayons correspondingly when reading the teacher directions.

### English Language Development Leveled Activities

| Emerging Level | Expanding Level | Bridging Level |
|---|---|---|
| **Build Background Knowledge** Display a number line to 10. Write *count on* on the board. Point to each number as you count to four, and then leave your finger on the four. Say, *Now I will count on from four.* Count on three more from four. Say, *I counted on three more from four. What number is it?* Have students answer chorally or with a gesture. **seven** Repeat the activity with another addition problem. | **Public Speaking Norms** Display a number line. Say, *To find the sum of 2 + 5, I started with the greater addend five, and then I counted on two more. The answer is seven.* Emphasize /d/ at the end of the past tense words. Have a volunteer solve a similar problem, and then use the sentence frame: **I started with _____, and then I counted on _____ more. The answer is _____.** Ensure students emphasize /d/ for past tense. | **Cooperative Learning** Divide students into groups of three. Give each group a number line and 12–15 color tiles or buttons. Say, *Write an addition sentence using the tiles. Use your number line to count on and find the answer.* Give students several minutes to complete the task. Then have one student in each group share the problem they created and explain how they used *counting on* to find the answer. |

### Multicultural Teacher Tip

Individual praise and standing out from the crowd are frowned upon in some cultures. Students from these cultures may act embarrassed or uncomfortable when invited to the board to solve a problem or asked to share an answer with the class. They may be more comfortable during group classroom activities, and may prefer others in the group to share answers or demonstrate problem solving.

NAME _____ DATE _____

# Lesson 2 Concept Web
## *Count On to Add*

Use the number line to show how to count on. Start with the greater addend.

Start at 8. Count on 2.
8 + 2 = __10__

Start at 7. Count on 2.
7 + 2 = __9__

Start at 4. Count on 1.
1 + 4 = __5__

Start at 5. Count on 4.
5 + 4 = __9__

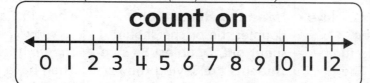

**count on**

0 1 2 3 4 5 6 7 8 9 10 11 12

3 + 6 = ?
Start at __6__.
Count on __3__.
3 + 6 = __9__

6 + 4 = ?
Start at __6__.
Count on __4__.
6 + 4 = __10__

2 + 7 = ?
Start at __7__. Count on __2__.
2 + 7 = __9__

5 + 6 = ?
Start at __6__. Count on __5__.
5 + 6 = __11__

**Teacher Directions:** Provide a description, explanation, or example of the new term by counting real objects and placing them on a number line. Have students say the term. Work the first item together. Have students practice saying the process using these sentence frames: **I start at eight and count on two. Eight plus two equals ten.** After students complete all items, have partners describe the process to each other.

# Lesson 3 Doubles and Near Doubles
## English Learner Instructional Strategy

### Vocabulary Support: Signal Phrases

Write *doubles* and its Spanish cognate, *dobles*, on a cognate chart. Introduce the term, and model with a math example. For See and Show, discuss that the kittens in the picture are the *same* just like the addends 6 and 6 are the *same number*. For near doubles in Exercises 2 and 5 stress the phrase *one more*. For Problem Solving Exercises, have pairs of students highlight the key signal phrases *same number* and *one more* before solving the problems.

### English Language Development Leveled Activities

| Emerging Level | Expanding Level | Bridging Level |
|---|---|---|
| **Number Game** Use index cards to assign pairs of students a number between 0 and 9. Say, *Who has the same number as you? Find your double.* Have students identify their partner according to their matching numbers. Then have pairs take turns coming to the board. Use each pair's numbers to demonstrate a doubles fact. For example, write on the board 3 + 3 = ____. Give students a chance to respond, either orally or with a gesture. Then say, *The sum is 6.* | **Building Oral Language** Give each student a pair of number cubes. Say, *On the count of three, roll your number cubes. One, two, three.* If a student rolls doubles, he or she should shout out, **Doubles!** Be sure students are saying the /z/ sound clearly to indicate the plural. Then use the double rolled to demonstrate a doubles fact and a near doubles fact. Repeat as time allows. | **Exploring Language Structures** Pair students, and give each pair 20 buttons. Assign each pair a number between 1 and 10. Say, *Use the buttons to show the doubles fact for the number you were given.* Then have each student use the following sentence frame to describe what they did: **When we doubled ____, the sum was ____.** Then have students model and describe a near doubles fact for their assigned number. Have students record their sentence frames and results in their math journals. |

**Teacher Notes:**

NAME _____ DATE _____

# Lesson 3 Word Web
## *Doubles and Near Doubles*

Draw a picture in each rectangle that shows the meaning of the math words.

See students' examples.

See students' examples.

**doubles**

**near doubles**

See students' examples.

See students' examples.

**Teacher Directions:** Provide a description, explanation, or example of the new terms using images or real objects. Have students say the terms aloud. Direct students to draw or write math examples that represent each of the math terms. Then encourage students to describe their pictures to a peer.

**4** Grade 2 • Chapter 1 *Apply Addition and Subtraction Concepts*

# Lesson 4 Make a 10

## English Learner Instructional Strategy

### Sensory Support: Manipulatives

Before the lesson, have a native language peer or mentor work with emerging/expanding students to introduce them to Work Mat 2 used in the Make a 10 lesson. Say, *This is a tool that makes it easier to see how objects add up to 10. We use counters in the boxes to show math problems.* Invite students to count the number of boxes in their native languages. Then have them count chorally to 10 and then to 20 in English.

Give each student Work Mat 2 and 20 counters. Have them experiment with different ways to make 10, writing their results in their math journals. Encourage them to count in their native languages as needed before counting in English.

### English Language Development Leveled Activities

| Emerging Level | Expanding Level | Bridging Level |
|---|---|---|
| **Choral Responses** | **Phonemic Awareness** | **Making Connections** |
| Pair students. Give one student 6 color tiles and the other student 5 tiles. Say, *She has six. How many more to make ten?* Give students a moment to answer chorally or with a gesture. Then say, *She needs four more to make ten.* Take 4 tiles from the second student and hand them to the first student. Say, *She has ten. How many tiles do they have in all?* Give students a moment to answer chorally or with a gesture. Then say, *Ten plus one makes eleven.* Repeat the activity as time allows. | Say, *I want to add seven and five. First I will make ten.* Emphasize the /k/ sound in make. Demonstrate solving the problem with color tiles. Say, *I made ten. Now I have ten and two. Ten plus two is twelve.* Emphasize the /d/ sound in *made.* Have students work in small groups to solve an addition problem using color tiles and the make ten strategy. Have them explain how they solved the problem, using the sentence frame: **First I made ten. Then I added ten and ____ to make ____.** Ensure students say the /d/ and /k/ sounds correctly. | Pair students. Ask each student to describe things they have made in the past, such as a paper airplane. Have them use the sentence frame: **Yesterday, I made ____.** Then have each student describe things they might like to make someday, using the sentence frame: **Someday, I will make ____.** Distribute 5 to 9 buttons to each student. Have students write an addition sentence based on the number of buttons each has and then record their answers in their math journals using the sentence frame: **____ and ____ make 10. 10 plus ____ makes ____.** |

### Teacher Notes:

NAME _____ DATE _____

# Lesson 4 Sum Identification
## *Make a 10*

Match addition examples to show how to make
a 10 to add.

= 10 + 2

8 + 4

= 10 + 4

= 10 + 1

9 + 5

= 10 + 3

Draw a picture to show one way to make a 10 to add.

See students' examples.

**Teacher Directions:** Model an addition sentence, such as 8 + 3, and use manipulatives to show how to make a 10 to find the sum. Teach, model, and prompt the sentence frames: **I have 8 and 3. I can take apart 3 to make a 10. Then I add.** Have students match an example on the left with the make-a-10 example on the right. Then direct students to draw a picture representing a way to make a 10 when adding. Encourage students to use the sentence frames above to describe their picture to a peer.

**Grade 2 • Chapter 1** *Apply Addition and Subtraction Concepts* **5**

# Lesson 5 Add Three Numbers
## English Learner Instructional Strategy

### Language Structure Support: Act It Out

During Explore and Explain, model using pictures of 4 fish, 6 frogs and 2 snakes as you present the teacher directions.

Before See and Show, invite 3 students to the front of class. Say, *This is a group.* Invite 2 more students and say, *This is another group. There are two groups.* Then invite 1 student and say, *This is another group. There are now three groups. Sometimes we use the word* **group** *to describe adding. Now I will group or add all the students together. First I will add 2 to 1 to make a group of 3. Now I have two groups of 3. These are doubles. 3 plus 3 equals 6.*

### English Language Development Leveled Activities

| Emerging Level | Expanding Level | Bridging Level |
|---|---|---|
| **Internalize Language** | **Phonemic Awareness** | **Making Connections** |
| Organize students into groups of three. Give students color tiles and have them imitate you as you model grouping. Say, *I have two groups of three tiles and a group of seven tiles. How can I group them together to add?* Give students a moment to group their tiles. Then have them answer chorally or with a gesture. Then say, *I have thirteen tiles. First, I grouped three and seven together to make ten. Then I added three.* Repeat the activity as time allows. | Use color tiles to model adding 3, 4, and 6 by grouping addends. Say, *First, I group two addends. I will group 4 and 6 to make 10.* Emphasize the /oo/ sound in *group*. Say, *Then I add 10 and 3 to get 13.* Give students a similar grouping problem to solve, such as 6 + 3 + 6. As students work, ask a volunteer to describe how he or she is using grouping to solve the problem. Provide a sentence frame, such as: **First, I grouped _____ and _____ to get _____. Then I added _____ and _____.** Be sure students are correctly saying the /oo/ sound in group. | Have students get into groups of three. On the board, write 4, 2, 4. Say, *Write a word problem that uses these addends.* Give groups several minutes to write their own word problems. Then say, *Solve your word problem. Show how you used grouping.* After students have found the solution, ask for volunteers to describe how they solved the problem by grouping. Discuss different ways to group the three numbers; looking for a known fact, grouping doubles first, or making a ten. |

### Teacher Notes:

NAME _____ DATE _____

# Lesson 5 Note Taking
## *Add Three Numbers*

Read the question. Write words you need help with. Use your lesson to write your Cornell notes.

| **Building on the Essential Question** | **Notes:** |
|---|---|
| How can I use strategies to add three numbers? | I can group <u>addends</u> in different ways. <br><br> 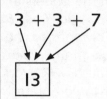 <br> 6 + 7 = ?  10 + 3 = ? <br><br> The sum will stay the <u>same</u>. <br><br> 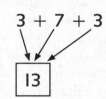 <br><br> I should look for <u>facts</u> I know. <br> I can make a <u>10</u>. <br> Add two numbers first. Then add the other number to find the <u>sum</u>. |
| **Words I need help with:** <br><br> *See students' words.* | |

**Teacher Directions:** Read the Building on the Essential Question and have students list words/phrases they need assistance with. Provide descriptions, explanations, or examples of the terms using images or real objects. Read each sentence frame and have students write the appropriate terms. Have students read their notes aloud.

# Lesson 6 Problem Solving
# Strategy: Write a Number Sentence
## English Learner Instructional Strategy

### Cooperative Learning: Partners Work/Pairs Check

Assign Apply the Strategy and Review the Strategies exercises to student pairs. One student completes the first exercise while the second acts as a coach. Then, students switch roles for the second exercise. When they finish the second exercise, they get together with another pair and check answers.

Provide the following sentence frames: **What number sentence did you get for Exercise \_\_\_\_? Our number sentence is \_\_\_\_.**

When both pairs have agreed on the answers, ask them to shake hands and continue working in original pairs on the next two exercises.

### English Language Development Leveled Activities

| Emerging Level | Expanding Level | Bridging Level |
|---|---|---|
| **Signal Words** | **Sentence Frames** | **Signal Words** |
| Use the Lesson 6 Reteach pages 1 and 2 to review the problem solving process. Read each word problem aloud for students and direct them to highlight key words/phrases such as: *total, each, in all,* and *more.* Then read each word problem again and invite students to echo read with you. Provide pictures of key objects: pencils, bottles, flowers, tricycles, and stickers. Guide students to underline what they know and circle what they need to find. Encourage them to solve on their own. | Draw four large boxes on the board, and label them Understand, Plan, Solve, and Check. In the Understand box, write *A bowl has 8 kiwis. I add 3 oranges to the bowl. How many pieces of fruit in the bowl?* Work together to complete the graphic organizer. Provide sentence frames for their answers, such as: **We know \_\_\_\_ and \_\_\_\_. We need to find \_\_\_\_. You should underline \_\_\_\_ You should circle \_\_\_\_** Ask questions about how they should solve the problem and how they can check their answer. | Present a demonstration problem solving organizer. In the Understand box, write: *The first part of our trip took 3 hours. The second part of the trip took 4 hours. How long was our trip?* Ask students to identify what they know, and then underline the first two sentences of the problem. On the board, write: *who, what, where when, how,* and *why. Say, These are question words that often tell you what you need to find. Which question word do you see in this problem?* **how** Have students use a graphic organizer to solve. |

### Teacher Notes:

NAME _____ DATE _____

# Lesson 6 Problem Solving
## *STRATEGY: Write a Number Sentence*

<u>Underline</u> what you know. (Circle) what you need
to find. Write a number sentence to solve.

**I.** <u>There are **3** rabbits at a farm.</u>

<u>The farmer buys **9 more**.</u>

(How many rabbits are there **now**?)

_3_ (+) _9_ (=) _12_

rabbit

| Part | Part |
|------|------|
|      |      |
| Whole ||
|      |      |

There are __12__ rabbits at the farm now.

---

**2.** **Chung** always **picks 5** flowers at home
after school.

(How many flowers will
he have after **3 days**?)

flower

_5_ (+) _5_ (+) _5_ (=) _15_

Chung will have __15__ flowers after
3 days.

**Teacher Directions:** Provide a description, explanation, or example of the bold face
terms and nouns using gestures, images, or real objects. Read each sentence and have
students echo read. Encourage students to write a number sentence and then write their
answers in each restated question. Have students read each answer sentence aloud.

Grade 2 • **Chapter I** *Apply Addition and Subtraction Concepts*  **7**

# Lesson 7 Count Back to Subtract
## *English Learner Instructional Strategy*

### Graphic Support: Number Lines

Before the lesson, draw a number line from 0 to 12 on a large sheet of chart paper and place it on the floor. Provide students with a sentence strip to make their own number line or prepare strips ahead of time. Have students work in pairs and listen to one another as they take turns counting forward to 20. Next, have them count back from 12 together using their number lines. Pair ELs with a native English speaker and have them walk backward on the floor number line counting backward together as they step on each number. Allow students to use their number lines or act out the exercises on the floor number line as they work through the student pages.

### English Language Development Leveled Activities

| Emerging Level | Expanding Level | Bridging Level |
|---|---|---|
| **Basic Vocabulary**<br><br>Ask, *Where is my back?* Give students a moment to answer, either orally or with a gesture. Then point to your back, and say, *back.* Walk backwards as you say, *I am moving back.* Show a number line on the board. Point to 10, then count back 6 places on the number line. Say, *Ten. I am moving back. 1, 2, 3, 4, 5, 6. I began at 10. I counted back 6. Now I am on 4. The difference is 4.* Have volunteers come to the number line. Say, *Show me _____ (number). Count back _____ (number). What is the difference?* Help students complete the task. | **Exploring Language Structure**<br><br>Use masking tape to create a number line on the floor. Have a volunteer stand on the 10. Say, *Start on 10. Take 9 steps back. Where are you?* 1 Write the equation, $10 - 9 =$ _____. Then ask the student how he or she found the difference. Post and model the following sentence frames as a communication guide: **I started on _____. I counted back _____. Then I was on _____. The difference is _____.** Invite volunteers to repeat the activity with new problems as time allows. | **Pairs Share**<br><br>Describe solving an equation by counting back. Have students get into pairs. Give each pair a number line to 12 and a number cube. Randomly assign pairs numbers from 0 to 12. Say, *Roll your number cube. Subtract the number you roll from the number you were assigned. Count back to find the answer.* Give students time to complete the task. Then have a student from each pair describe the process. For example, **We were given the number 12. We rolled 3. We counted back 3 from 12. 12 take away 3 is 9. The difference is 9.** |

### Teacher Notes:

NAME _____ DATE _____

# Lesson 7 Vocabulary Sentence Frames
## *Count Back to Subtract*

The math words in the box are for the sentences below. Write the words that fit in each sentence on the blank lines.

| count back | difference | subtract |
|---|---|---|

1. The ___difference___ is the answer to a subtraction problem.

$$6 - 2 = 4$$

2. To ___subtract___ means to take away, take apart, separate, or find the difference between two sets.

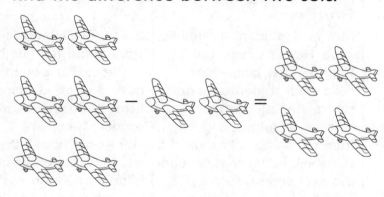

3. On a number line, start at the greater number and ___count___ ___back___ if you want to subtract.

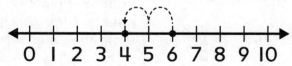

0 1 2 3 4 5 6 7 8 9 10

 **Teacher Directions:** Provide a description, explanation, or example of the each term using images or real objects. Read each sentence frame and have students echo read. Direct students to write the correct terms in each blank. Then encourage students to read each sentence to a peer.

# Lesson 8 Subtract All and Subtract Zero

## English Learner Instructional Strategy

### Vocabulary Support: Cognate

Before the lesson, write *zero* and its Spanish cognate, *cero,* on a cognate chart. Introduce the review vocabulary word, write a math example, and utilize concrete objects to model subtracting all from a group to end up with none or zero.

Have students use an index card to write the term and draw a visual example. Next, have students write a definition and sentence using *zero* in context. Provide examples for each on the chart. For non-Spanish speaking ELs, refer to the Multilingual eGlossary, or accommodate the student with a translation tool. Have students glue a pocket into their math journals to store the cognate card for future review.

### English Language Development Leveled Activities

| Emerging Level | Expanding Level | Bridging Level |
|---|---|---|
| **Act It Out** Invite 3 students to stand in front of the board. On the board above them write $3 - 3 =$ _____. Have students act out the problem. As you say, *3 students take away 3 students equals,* have the three students sit back down. Ask, *How many students are left?* Give students a moment to answer chorally or with a gesture, **zero.** As you say, *No students are left. There are 0 students,* write 0 on the board to complete the number sentence. Repeat the activity with $3 - 0 =$ _____ to model subtracting 0. | **Exploring Language Structure** Give each student a number cube. Have students roll their number cube. Then have each student describe subtracting all of the number or subtracting zero from the number. Provide sentence frames that contain a causal phrase, such as: **If I take away ____ from ____, then I will have zero,** or **If I take away 0 from ____, then I will have ____.** | **Listen, Write, and Read** Write $3 - 0 = 3$ on the board. Tell a brief number story about the number sentence. For example, **I have 3 dogs at home. I bring 0 dogs with me to school. There are 3 dogs still at my home.** Divide students into groups of 3. Randomly assign each group a number from 1 to 10. Have groups write number stories that subtract all of their number or subtract zero from their number. Have a volunteer from each group share their number story. |

**Teacher Notes:**

NAME _____ DATE _____

# Lesson 8 Note Taking
## *Subtract All and Subtract Zero*

Read the question. Write words you need help with. Use your lesson to write your Cornell notes. Write or draw math examples to explain your thinking. Share your examples with a classmate.

| | |
|---|---|
| **Building on the Essential Question**<br><br>How can I subtract 0?<br>How can I subtract all? | **Notes:**<br><br>I know that __subtract__ means "take away."<br><br>The answer to a subtraction problem is the __difference__.<br><br>If I subtract __all__, the difference is 0. |

**Notes (continued):**

If I subtract __all__, the difference is 0.

$$\cancel{} \cancel{} \cancel{} \cancel{} - \cancel{} \cancel{} \cancel{} \cancel{} = 0.$$

**Words I need help with:**

See students' words.

If I subtract 0 from a number, the difference is the __same__ as that number.

$$\cancel{} \cancel{} - 0 = \cancel{} \cancel{}$$

**My Math Examples:**

See students' examples.

**Teacher Directions:** Read the Building on the Essential Question and have students list words/phrases they need assistance with. Provide descriptions, explanations, or examples of the terms using images or real objects. Read each sentence frame and have students write the appropriate terms. Have students read their notes aloud. Direct students to create an example representing the question. Then encourage students to describe their picture to a peer.

**Grade 2 • Chapter 1** *Apply Addition and Subtraction Concepts* **9**

# Lesson 9 Use Doubles to Subtract

## English Learner Instructional Strategy

### Sensory Support: Realia

As an introduction to the lesson, model the word *doubles* using gestures, facial expressions, and props. Provide pictures for the key objects; *pie, cherry, pumpkin,* and *farmers' market* to visually support the Problem Solving exercises. Ask the following tiered questions for their formative assessment.

For emerging level students, ask yes/no questions utilizing lesson vocabulary such as, *Are these doubles facts?* Look for understanding through demonstration, nodding, and/or pointing.

For expanding and bridging level students, ask questions that encourage a simple sentence response such as, *How did you get this answer?* and *Why is this problem a doubles fact?*

### English Language Development Leveled Activities

| Emerging Level | Expanding Level | Bridging Level |
|---|---|---|
| **Cognates**<br>Write *difference* and its Spanish cognate, *diferencia,* on a cognate chart and on the board. Then write 4 − 2 = 2 next to *difference.* Circle the 2s and point to them as you say, *doubles fact.* Point to the word *difference* on the board. Discuss the definition using the *difference* My Vocabulary Card for reference. Have a volunteer draw a line from the word *difference* to the difference in the number sentence. Have the student circle the difference. Repeat the activity with a new volunteer and a new number sentence. | **Sentence Frames**<br>Pair students. Give each student 1 to 10 connecting cubes of a single color. Make sure partners have equal numbers of cubes but of different colors. Have each pair connect their cubes to show a doubles addition fact and describe it. Provide a sentence frame, such as: \_\_\_\_ **cubes plus** \_\_\_\_ **cubes is** \_\_\_\_ **cubes in all.** Then have each pair use their cubes to describe a doubles subtraction fact. Provide a sentence frame, such as: \_\_\_\_ **cubes take away** \_\_\_\_ **cubes leaves** \_\_\_\_ **cubes.** | **Exploring Language Structure**<br>Ask each student to write three doubles subtraction problems. For example, 12 − 6 = 6, 4 − 2 = 2, and 10 − 5 = 5. Have students exchange problems. Ask each student to describe the problems using the words *different* and *difference.* For example: **I have three different subtraction problems. The difference of 10 minus 5 is 5.** |

### Teacher Notes:

NAME _____ DATE _____

## Lesson 9 Four-Square Vocabulary

*Use Doubles to Subtract*

Write the definition for *doubles*. Write what the word means, draw a picture, and write your own sentence using the word.

| **Definition** | **My Own Words** |
|---|---|
| Two addends that are the same number. | See students' examples. |
| **My Picture** | **My Sentence** |
| See students' drawings. | 5 + 5 is a doubles fact. I can use that to help me subtract 10 − 5. |

**doubles**

**Teacher Directions:** Provide a description, explanation, or example of the new term using images or real objects. Have students use the Glossary to write the definition. Direct students to write a definition in their own words and draw a picture representing their math term. Have students write a sentence using the term. Then encourage students to read their sentence to a peer.

# Lesson 10 Relate Addition and Subtraction

## English Learner Instructional Strategy

### Sensory Support: Pictures and Photographs

ELs may have trouble with Problem Solving Exercises if they do not understand the key nouns and verbs being discussed. Provide picture and/or photographic support whenever possible. In this lesson, students may not know the following English terms: *tennis, team, players, turtles, water, swim, log, fly(ing)* or *birds.* As they work through the problems, it may be helpful for students to draw pictures or use symbols to represent the items in each problem.

Have students highlight key subtraction words/phrases: *leave, are still, away,* and *are left,* in word problems. Create a classroom subtraction word list for students to reference.

### English Language Development Leveled Activities

| Emerging Level | Expanding Level | Bridging Level |
|---|---|---|
| **Build Word Knowledge** Write *related* on the board and say the word. Then write *family*, and below it write *mom, dad, brother,* and *sister.* Be sensitive to other groupings. Say, *These are people in a family. They are related.* Write $5 + 4 = 9$ on the board, and then below it, write $9 - 4 = 5$. Draw lines to connect each number in the top equation to the same number in the bottom equation. Say, *These two problems are related.* Repeat with new number sentences. Have a volunteer draw lines connecting the matching numbers in the equations. | **Building Oral Language** On the board, write $7 + 8 = 15$. Below the equation write, $15 - 7 = 8$ and $15 - 8 = 7$. Say, *If I know $7 + 8 = 15$, then I also know $15 - 7 = 8$ and $15 - 8 = 7$.* Write addition sentences on index cards, and distribute one to each student. Have students use the equations on their cards to describe related facts. Provide a causal sentence frame, such as: **If I know _____, then I also know _____.** If students describe only one related fact, prompt then to describe another one by asking, *What else do you know?* | **Number Game** Have students form two lines. Invite the two students at the front of the lines to come to the board. Write an addition sentence, such as, $8 + 4 = 12$. Then say, *Write a related fact.* The first student to finish writing a related fact and describing the number sentence correctly goes to the back of the line. The other student takes a seat. If there is a tie, both students go back and stay in line. Have the next two students come forward. Write a new addition sentence and continue until all students have had a turn or until one student remains. |

## Teacher Notes:

NAME _____ DATE _____

# Lesson 10 Note Taking

## *Relate Addition and Subtraction*

Read the question. Write words you need help with. Use your lesson to write your Cornell notes. Write or draw math examples to explain your thinking. Share your examples with a classmate.

| **Building on the Essential Question** | **Notes:** |
|---|---|
| How can I relate addition to subtraction? | I can use __addition__ facts to help me subtract. |
| | __Related__ __facts__ have the same three numbers. |
| | You can write related __addition__ and __subtraction__ facts. |
| | $3 + 6 = 9$ |
| | $6 + 3 = 9$ |
| | $9 - 3 = 6$ |
| | $9 - 6 = 3$ |
| **Words I need help with:** | Addition and subtraction are opposite or __inverse__ operations. |
| See students' words. | You can use $3 + 6 = 9$ to find $9 - 3 = $ __6__. |

**My Math Examples:**

See students' examples.

**Teacher Directions:** Read the Building on the Essential Question and have students list words/phrases they need assistance with. Provide descriptions, explanations, or examples of the terms using images or real objects. Read each sentence frame and have students write the appropriate terms. Have students read their notes aloud. Direct students to create an example representing the question. Then encourage students to describe their picture to a peer.

# Lesson 11 Missing Addends

## English Learner Instructional Strategy

### Vocabulary Support: Frontload Academic Vocabulary

Before the lesson, write *addend, related facts,* and *missing addend* on a classroom chart. Introduce the words, write a math example, and provide concrete objects to support understanding. Assist non-Spanish speaking ELs with an appropriate translation tool. Have students write the terms in both English and their first language in a math journal and draw a visual example for each. Next, have multilingual pairs write definitions and sentences using each term in context. Have pairs place an addition number sentence example on the chart and then report back using specific vocabulary.

### English Language Development Leveled Activities

| Emerging Level | Expanding Level | Bridging Level |
|---|---|---|
| **Activate Prior Knowledge** | **Report Back** | **Word Knowledge** |
| On the board, write the letters A through J, but leave a blank space where D should be. Point to the space, and say, *One letter is not here. It is missing. What letter is missing?* Give students a chance to answer **D**, then say, *Yes, the letter D is missing.* Write D in the blank space. On the board, write 5 + ___ = 8. Point to the space, and say, *This addend is missing.* Model finding the missing addend using subtraction. Repeat providing addition number sentences for students to find missing addends on their own. Have pairs compare answers. | Have students get into groups of three. Say, *Write an addition sentence with a sum less than 20. Then, on a different sheet of paper, rewrite the problem with one of the addends missing.* If necessary, model the task. Have groups exchange problems and use subtraction to find the missing addend. Have one student from each group identify the missing addend and explain their reasoning. Provide a sentence frame, such as: **I know the missing addend is ___ because ___.** | On the board, write: *first, second, next, then,* and *last.* Say, *These are sequence words. They are used to tell the order of events, or the order that something happens.* On index cards, write addition sentences with missing addends, such as, 3 + ___ = 9. Give each student a card. Ask volunteers to use the sequence words to describe how they found the missing addend. For example: **First, I wrote a subtraction problem. Then I solved it. The difference was the missing addend. Last, I checked my answer by using the addend in the addition problem.** |

**Teacher Notes:**

NAME _____ DATE _____

# Lesson II Vocabulary Definition Map
## *Missing Addends*

Use the definition map to write what the math word means and tell what the word is like. Write or draw a math example. Share your examples with a classmate.

**My Math Words:**

missing addend

**What It Is Like:**

__Addends__ are the numbers you add together.

**What It Means:**

The missing number in a number sentence that makes the number sentence true.

$5 + \square = 9$
The $\square$ is the __missing__ addend.

Use a __related__ __fact__ to help find the missing addend.

**My Math Example:**

See students' examples.

**Teacher Directions:** Provide a description, explanation, or example of the new term using images or real objects. Have students use the lesson or Glossary to define the math term. Direct students to list characteristics, and draw a picture representing their math term. Then encourage students to describe their picture to a peer.

# Lesson 12 Fact Families

## English Learner Instructional Strategy

### Language Support: Echo Reading

Assign Problem Solving Exercises 7–9. Have students work in pairs. Have one student read the first problem sentence-by-sentence while the partner listens and then echo reads the problem back sentence-by- sentence. Then, have them work together to solve the problem. Have students switch roles for the next problem. When they finish all three problems, they report back to the teacher using the following sentence frame:

For Exercise _____ (7/8/9) there are _____ animals/apples/children in all.

During echo reading, emerging level students may benefit from working with the teacher or a native English speaker.

### English Language Development Leveled Activities

| Emerging Level | Expanding Level | Bridging Level |
|---|---|---|
| **Basic Vocabulary** | **Phonemic Awareness** | **Cooperative Learning** |
| Write the term *fact family.* Show photos of families. As you show each photo, say, *This is a family. The people in a family are related.* Say, *A fact is true.* Give several examples of facts and identify them as such. For example, *My name is _____. That is a fact.* Finally, write $4 + 2 = 6$ on the board, and say, *Four plus two is six. That is a fact.* Then write the remaining number sentences of the fact family, and say, *These are also facts. They are all related. They are called a fact family.* Guide students to refer to the glossary to read the definition of *fact family.* | Write the fact family for 4, 5, and 9 on the board. Say, *These are all facts. They are related. They make up a fact family.* Emphasize the final /kt/ sound of *fact.* Write *fact family* on the board. Write another fact family on the board. Ask, *Is this a fact family? How do you know?* Give students a chance to answer using complete sentences. Repeat with a set of number sentences that are not a fact family, and have students explain why it is not a fact family. Be sure students are saying the final /kt/ sound of *fact.* | Write the four equations of a fact family on separate index cards. Create several sets for different fact families, enough so there is one card for each student. Randomly distribute the cards to students. Say, *Find the members of your fact family.* Give students several minutes to identify their groups according to the number sentences on their cards. Then give each group a sheet of paper and crayons or colored pencils. Say, *Draw a picture that shows your fact family.* Afterward, have groups explain their drawings. |

## Teacher Notes:

NAME _____  DATE _____

# Lesson 12 Concept Web
## *Fact Families*

Draw lines to match the items to the term *fact family*.

$2 + 5 = 7$
$5 + 2 = 7$
$7 - 2 = 5$
$7 - 5 = 2$

$5 + 4 = 9$
$5 - 4 = 1$
$9 + 4 = 13$
$9 + 5 = 13$

$6 - 5 = 1$
$5 - 4 = 1$
$4 - 3 = 1$
$3 - 2 = 1$

**fact family**

$6 + 6 = 12$
$12 - 6 = 6$

$9 + 8 = 17$
$8 + 9 = 17$
$17 - 8 = 9$
$17 - 9 = 8$

$9 + 3 = 12$
$3 + 9 = 12$
$12 - 9 = 3$
$12 - 3 = 9$

**Teacher Directions:** Provide a description, explanation, or example of the new term using images or real objects. Have students say the math term aloud. Direct students to draw a line from each example of a fact family to the term in the center. Then encourage students to explain why two ovals do not contain a fact family to a peer.

**Grade 2 · Chapter 1** *Apply Addition and Subtraction Concepts*   **13**

# Lesson 13 Two-Step Word Problems

## English Learner Instructional Strategy

### Language Support: Tiered Questions

As students complete the Talk Math activity, provide sentence frames to discuss the prompt. Have students explain how to solve a two-step word problem in Exercise 1 or 2. Post the sentence frames in a readily visible location. Emerging ELs will benefit from a simple question, such as: *Does this problem require one step or two steps?* **2 steps** Encourage Expanding/Bridging level students to answer the prompt using more challenging sentence frames: **You solve a two-step problem by answering the _____ (first) part. Then you solve the _____ (second) part. I solved the first/second part by _____ (adding/ subtracting) _____ and _____.** Have all students finish with: **The answer is _____.**

### English Language Development Leveled Activities

| Emerging Level | Expanding Level | Bridging Level |
|---|---|---|
| **Exploring Language Structures** Write *step* and *steps* on the board. Take a step, and say, *one step.* Then take two steps, and say, *two steps.* Emphasize the final /s/ as you point to the final s of *steps* on the board. Say, *5 cats on a fence. 3 more cats get on the fence. Then 1 cat leaves. How many cats?* Write $5 + 3 = 8$ on the board. Point to the equation, and say, *one step.* Then write $8 - 1 = 7$ on the board. Say, *second step.* Ask, *How many steps?* Give students a chance to answer **2**, then say, *two steps.* Emphasize the final /s/. Explain that some word problems take two steps to solve. | **Multiple Word Meanings** Write *steps* on the board. Ask students to describe *steps.* Ensure they reference walking or stairs. Remind students that some words have more than one meaning. Say, *Sometimes you need to use two steps to solve a word problem.* On the board, write: *8 fish are in a pond. 6 fish join them. Then 4 fish leave. How many fish in the pond?* Model the problem. Say, *The first step was addition. The second step was subtraction. Now we know the answer is 10 fish.* | **Signal Words** On the board, write: *first, next,* and *last.* Say, *These are sequence words. They are used to tell the order of when things happen.* On the board, write the equations: $4 + 8 = 12$ and $12 - 2 = 10$. Pair students. Say, *Tell a number story using sequence words and the equations on the board.* Give students several minutes to write their stories. If necessary, provide sentence frames, such as: **First 4 _____. Then 8 _____. Last 2 _____, so 10 _____.** Ask pairs to share their stories with the other students. |

## Teacher Notes:

NAME _____ DATE _____

# Lesson 13 Guided Writing
## *Two-Step Word Problems*

**How do you solve a two-step word problem?**

Use the exercises below to help you build on answering the Essential Question. Write the correct word from the word bank on the lines provided.

**1.** What key words do you see in the question?

two-step, word problem
_____

marbles

| Camden has 3 marbles in his pocket. He finds 8 more. He gives 4 marbles to Jen. How many marbles does Camden have now? | **Word Bank**<br>first<br>know<br>read<br>solve |
| --- | --- |

**2.** First, I should ___read___ the word problem.
I should underline what I ___know___ .

**3.** Next, I decide what part to solve ___first___ .

**4.** Then use the rest of the information to ___solve___ the next part.

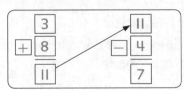

**5.** How do you solve a two-step word problem?

I would use the information I have, decide which part to solve first,
_____

and then use the rest of the information to solve the next part.
_____

 **Teacher Directions:** Read the Building on the Essential Question. Provide descriptions, explanations, or examples of the terms using images or real objects. Read each sentence frame and have students use the lesson and Glossary to write the appropriate terms. Have students read the sentences aloud.

# Chapter 2 Number Patterns

## What's the Math in This Chapter?

### Mathematical Practice 7: Look for and make use of structure

Distribute a hundreds chart and 20 connecting cubes to each student. Say, *We need to count these really fast. Let's put these in groups so we can count them at the same time.* Model putting two connecting cubes together. Repeat until all 20 connecting cubes are in groups of 2. Say, *Now put your connecting cubes into groups of two.*

Have students chorally count by 2s with you using their groups of 2 connecting cubes. Then have them color even numbers 2–20 on their hundreds charts. Discuss any patterns that students see in the hundreds chart and encourage them to continue the coloring pattern. Skip count even numbers with students, starting with 0 and clapping on odd numbers (0, clap, 2, clap, etc.).

Discuss how skip-counting is "skipping" a number because we already know how many are in a group. Show 3 groups of unequal connecting cubes: 2, 5 and 3. Ask, *Could we skip count if the groups are **not** equal?* **No, because we cannot use a pattern.** The goal of the discussion is to highlight how skip counting involves using structure/patterns to count and can be used to find the sum in a repeated addition problem.

Display a chart with Mathematical Practice 7. Restate Mathematical Practice 7 and have students assist in rewriting it as an "I can" statement, for example: **I can see how using equal groups and patterns helps me understand numbers.** Post the new "I can" statement.

## Inquiry of the Essential Question:

### How can equal groups help me add?

Inquiry Activity Target: **Students come to a conclusion that they can look for equal groups to help with problem solving.**

As an introduction to the chapter, present the Essential Question to students. The inquiry graphic organizer will offer opportunities for students to observe, make inferences, and apply prior knowledge of equal groups representing the Essential Question. As they investigate, encourage students to draw, write, and collaborate with peers to demonstrate their observations and thinking. Then have students present additional questions they may have to a peer to extend discussions.

Regroup students and restate Mathematical Practice 7 and the Essential Question. Pose questions to reflect on what has been learned to guide students in making connections between the Mathematical Practice and the Essential Question.

NAME _____ DATE _____

# Chapter 2 Number Patterns

## *Inquiry of the Essential Question:*

### How can equal groups help me add?

Read the Essential Question. Describe your observations
(I see...), inferences (I think...), and prior knowledge (I know...)
of each math example. Write any questions you have.

$5 + 5 + 5 = 15$

I see ...

I think ...

I know ...

| 1 | 2 | 3 | 4 | 5 | 6 | 7 | 8 | 9 | 10 |
|----|----|----|----|----|----|----|----|----|----|
| 11 | 12 | 13 | 14 | 15 | 16 | 17 | 18 | 19 | 20 |
| 21 | 22 | 23 | 24 | 25 | 26 | 27 | 28 | 29 | 30 |
| 31 | 32 | 33 | 34 | 35 | 36 | 37 | 38 | 39 | 40 |
| 41 | 42 | 43 | 44 | 45 | 46 | 47 | 48 | 49 | 50 |
| 51 | 52 | 53 | 54 | 55 | 56 | 57 | 58 | 59 | 60 |
| 61 | 62 | 63 | 64 | 65 | 66 | 67 | 68 | 69 | 70 |
| 71 | 72 | 73 | 74 | 75 | 76 | 77 | 78 | 79 | 80 |
| 81 | 82 | 83 | 84 | 85 | 86 | 87 | 88 | 89 | 90 |
| 91 | 92 | 93 | 94 | 95 | 96 | 97 | 98 | 99 | 100 |

I see ...

I think ...

I know ...

Write the missing numbers:

20, 30, ____, ____, 60, ____, ____, 90

Questions I have...

_____

_____

**Teacher Directions:** Read the Essential Question for students. Have students echo read.
Direct students to describe their observations, inferences, and prior knowledge of each
math example. Encourage students to write or draw additional questions they may have.
Then have students share their ideas/questions with a peer.

Grade 2 • **Chapter 2** *Number Patterns* **15**

# Lesson 1 Skip Count on a Hundred Chart

## English Learner Instructional Strategy

### Vocabulary Support: Math Word Wall

Before the lesson, write *skip count* on a chart. Discuss the term, write a math example, and provide concrete objects to support understanding. Review the definition in the glossary and add it to the Math Word Wall. Pair ELs with native English-speaking peers. Have them count together from 1 to 100 to practice counting before the lesson. Then have them skip count by 2s from 2 to 50 either chorally together or using echo counting.

Reinforce counting skills by providing multiple opportunities for choral counting during the lesson.

## English Language Development Leveled Activities

| Emerging Level | Expanding Level | Bridging Level |
|---|---|---|
| **Exploring Language Structure** | **Phonemic Awareness** | **Number Game** |
| Have 8 students stand in a line. Say, *I will count these students.* Point to one student, and say, *But I will skip _____ (student's name).* Count the students but say, *skip* when it is time to count the student you named. Then say, *I counted the students, but I skipped _____*, emphasizing the ending sound /t/ in *skipped.* Model counting by 2s on a number line, drawing arcs as you count. Point to the odd numbers and ask, *Did I count these numbers or did I skip these numbers?* Give students a moment to answer, then say, *I skipped these numbers.* | Display a hundred chart. Say, *I am going to skip count by 5s.* Emphasize the /sk/ sound of *skip.* Start on 5 and skip count to 30, coloring in each number as you count it. Point to 33 and ask, *Do I count this number, or do I skip it?* Have students answer chorally, **Skip it!** Be sure they are saying the /sk/ sound clearly. Point to 35 and ask, *Count it or skip it?* Have students answer chorally. Continue skip counting to 100, stopping a few more times to ask students, *Count it or skip it?* Repeat the activity, skip counting by 2s and 10s. | Divide the class into 4 teams, and number them 1 to 4. Give each team a hundred chart. Say to Team 1, *Start on 35. Skip count by 5s three times. What is the number?* Give Team 1 a moment to tell you the answer. **50** Then say to Team 1, *Give Team 2 a skip-counting problem. You can ask them to skip count by 2s, 5s, or 10s.* Monitor the instructions groups provide and make suggestions as necessary. Once Team 2 has found the answer, have them provide a skip counting problem for Team 3, and so on. |

## Teacher Notes:

NAME _____ DATE _____

# Lesson 1 Concept Web
## *Skip Count on a Hundred Chart*

Use the word web to show examples of skip counting.

by 2s from 10:

10, __12__, __14__, __16__

by 5s from 25:

25, __30__, __35__, __40__

by 10s from 50:

50, __60__, __70__, __80__

## skip count

by 5s from 85:

85, __90__, __95__, __100__

by 5s from 40:

40, __45__, __50__, __55__

by 10s from 30:

30, __40__, __50__, __60__

**Teacher Directions:** Provide a description, explanation, or example of the new term. Practice skip counting by 2s, 5s, and 10s with students. Use a 100 chart if necessary. After students have completed the web, have them orally compare answers with a partner.

# Lesson 2 Skip Count by 2s, 5s, and 10s
## English Learner Instructional Strategy

### Sensory Support: Pictures/Photographs

To introduce the lesson, use pictures of items in groups of 2, 5, and 10 to practice counting as a class. Use pictures of the items named in the Problem Solving exercises such as: bunches of grapes, shells, cookies and tennis balls, to provide additional language support.

Emerging level students may struggle with skip counting. Pair them with a peer or guide them in using pennies, nickels, and dimes to practice skip counting before the lesson.

### English Language Development Leveled Activities

| Emerging Level | Expanding Level | Bridging Level |
|---|---|---|
| **Listen and Identify** | **Memory Device/Mnemonic** | **Cooperative Learning** |
| Display a number line and a hundred chart. Model skip counting. Invite a student to come to the chart. Skip count by 5s up to 45, then point to 49 and ask, *Do I skip this number?* And then point to 50 and ask, *Or do I skip this number?* Give the student a chance to answer orally or with a gesture. Then say, *I skip 49 and I count 50.* Continue skip counting to 75, then ask, *Which number comes next?* Give the students a chance to answer, and then point to and say, *80.* Repeat the activity with another student, skip counting by 2s, 5s, or 10s. | Display a hundred chart and use a rhyming chant to help students skip count by 2s. Say, *Here's what I do when I count by two.* Have students repeat the chant several times. Then, as a group, skip count by 2s to 20. Repeat the activity with rhyming chants for skip counting by 5s and 10s. *I feel alive when I count by five. I point with a pen when I count by ten.* Using blocks or other manipulatives, display several groups of 2, 5, or 10 objects in each. Have students use the rhyming chants and skip counting to find the total number of objects in all the groups. | In multilingual groups, have students brainstorm chants that use skip counting. They may create their own rhymes or give examples that they have heard, such as **2, 4, 6, 8, who do we appreciate?** Have groups take turns presenting their chants to the class. Remind students to speak clearly and look at their audience. |

### Teacher Notes:

NAME _____ DATE _____

# Lesson 2 Vocabulary Definition Map
## Skip Count by 2s, 5s, and 10s

Use the definition map to write what the math word means and tell what the word is like. Write or draw a math example. Share your examples with a classmate.

**My Math Word:**

> ## equal groups

**What It Is Like:**

There are __4__ equal groups of counters.

**What It Means:**

Each group has the same number of objects.

When there are equal groups, __skip__ __count__ to find the total.

2, __4__, __6__, __8__

**My Math Example:**

See students' examples.

**Teacher Directions:** Provide a description, explanation, or example of the new term using images or real objects. Have students use the lesson or Glossary to define the math term. Direct students to list characteristics, and draw a picture representing their math term. Then encourage students to describe their picture to a peer.

**Grade 2 • Chapter 2** *Number Patterns* **17**

# Lesson 3 Problem Solving Strategy: Find a Pattern

## English Learner Instructional Strategy

### Collaborative Support: Act It Out

Frontload any vocabulary that may be unfamiliar to students such as: *favors, ticket, museum, map* (with map key showing miles), *stack*, and *food drive*, using pictures, and descriptions. Invite Expanding and Bridging level ELs to explain the words they know to peers that are struggling. Act out the Problem Solving exercise using bags and party favors. Put 5 party favors in each bag all at once while counting by 5s. Invite volunteers to come up and repeat the activity.

### English Language Development Leveled Activities

| Emerging Level | Expanding Level | Bridging Level |
|---|---|---|
| **Build Background Knowledge** Write and read aloud, *A pet store has 8 fish tanks. There are 5 fish in each tank. How many fish are there in all?* Provide pictures of nouns and define. Ask, *Do we know how many fish tanks there are? How many fish are in each tank?* Give students a chance to answer, then underline as you say, *Yes, we know there are 8 tanks and 5 fish are in each tank.* Then say, *A question mark tells me what we need to find. I will circle what we need to find.* Circle the question and say, *We do not know how many fish in all.* Stress *not.* Continue modeling to solve. | **Partner Work** Write and read aloud the following: *A roller coaster has 9 cars. 2 people sit in each car. How many people can ride the roller coaster?* Write Understand, Plan, Solve, and Check on four index cards with explanations for each step. Ask a pair of volunteers to come to the board. Hand the Understand card to one student. He or she will read it aloud as the second student follows the directions. Have additional pairs of volunteers come to the board to read and perform the other three steps in the problem-solving process. | **Share What You Know** Draw a large problem solving graphic organizer with four large boxes labeled: Understand, Plan, Solve, and Check. Divide students into four groups, and assign each group one of the problem-solving steps. On the board, write: *Each shelf holds 10 boxes. There are 8 shelves. How many boxes in all?* Then read the word problem aloud. Have the students in each group direct you in completing their assigned step. Guide students toward the correct skip counting pattern needed to solve the problem. Encourage use of a hundred chart if needed. |

**Teacher Notes:**

NAME _____ DATE _____

# Lesson 3 Problem Solving
## STRATEGY: Find a Pattern

<u>Underline</u> what you know. (Circle) what you need to find. Find a pattern to solve each problem.

**1. Xavier** is looking at a map of an **island**.

map

**He** (Xavier) knows that each finger **width** is about 10 **miles**.

How many miles will he count if he uses 7 finger widths?

10, _20_, _30_, _40_, _50_, _60_, _70_

finger

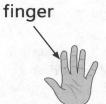

He will count _70_ miles.

---

**2. Sam** thinks of the **number pattern** 15, 20, 25, 30.

**He continues** this pattern.

What will be the next four numbers?

15, 20, 25, 30, _35_, _40_, _45_, _50_

The next four numbers will be _35_, _40_, _45_, _50_.

**Teacher Directions:** Provide a description, explanation, or example of the boldface terms and nouns using images or real objects. Read each sentence and have students echo read. Encourage students to find a pattern and then write their answers in each restated question. Have students read each answer sentence aloud.

# Lesson 4 Repeated Addition
## English Learner Instructional Strategy

### Graphic Support: Word Web

Before the lesson, write *repeated* and its Spanish cognate, *repetida*, on a cognate chart. Introduce the word, draw a math example, and provide concrete objects to support understanding. Create a word web with *repeated* in the center. Ask, *What have you repeated? Tell me about it.* Have them respond using the displayed sentence frame: **I have repeated _____.** Draw illustrations next to their examples on the word web. Then encourage Emerging/Expanding level ELs to answer verbally or by drawing a picture.

### English Language Development Leveled Activities

| Emerging Level | Expanding Level | Bridging Level |
|---|---|---|
| **Frontload Academic Vocabulary** Say, *Hi, my name is Mrs./Ms./Mr. _____.* Wait a moment, and then say it again. Say, *I said the same thing two times. I **repeated** what I said.* Emphasize *repeated.* Direct students to raise their hands and then lower them. Then have students repeat the movement. Say, *You raised your hands two times. You did it one time, and then you repeated it.* Write $3 + 3 + 3 = 9$. Ask, *What is repeated?* **3** Say, *This is called repeated addition.* | **Sentence Frames** Divide the class into four groups. Give each group a pair of (0–5) number cubes. Say, *Roll your cubes. Use the numbers to create a repeated addition problem. The bigger number is how many times you will add the smaller number.* Model the activity. After students have completed the activity, have groups tell what they did. Provide sentence frames, such as: **We rolled _____ and _____. We added _____ _____ times. We used repeated addition, and the total is _____.** | **Report Back** Write $4 + 4 + 4 + 4$ on the board. Have students get into pairs. Say, *Create a story for this repeated addition problem, and then solve it.* Have each pair share their story with the class. Be sure students are using complete sentences as they tell their stories. |

### Multicultural Teacher Tip

Some ELs may be more familiar with an educational system that emphasizes individualized work, learning from lectures, reading textbooks, and memorizing facts. As a result, they may be unfamiliar with an approach that stresses collaboration and problem solving. During group activities, be aware of students who are reluctant to communicate or participate. By prompting them with direct questions or by assigning them specific tasks, you can help the students take part in the group dynamic.

NAME _____ DATE _____

# Lesson 4 Vocabulary Definition Map
## *Repeated Addition*

Use the definition map to write what the math word means and tell what the word is like. Write or draw a math example. Share your examples with a classmate.

**My Math Words:**

repeated addition

**What It Is Like:**

Addends are numbers that are added together to find a sum.

**What It Means:**

To use the same addend over and over.

When groups are equal, use repeated addition to find the total.

You can skip count to find the sum.

**My Math Example:**

See students' examples.

**Teacher Directions:** Provide a description, explanation, or example of the new term using images or real objects. Have students use the lesson or Glossary to define the math term. Direct students to list characteristics, and draw a picture representing their math term. Then encourage students to describe their picture to a peer.

Grade 2 • **Chapter 2** *Number Patterns* **19**

# Lesson 5 Repeated Addition with Arrays

## English Learner Instructional Strategy

### Vocabulary Support: Draw Visual Examples

Before the lesson, pair ELs with an older peer or mentor. Write the words *array* and *row* on the board. Have the pairs work together to review the terms in both English and Spanish using the glossary. Then have students draw a visual example of an array in their math journals and label it using the glossary picture as a guide. They can then refer to their journals during the lesson to help remember the math terms.

For Exercise 10, remind students that the noun *wind turbines* was used in Lesson 4. Refer them back to see the picture.

### English Language Development Leveled Activities

| Emerging Level | Expanding Level | Bridging Level |
|---|---|---|
| **Word Recognition** | **Phonemic Awareness** | **Logical Reasoning** |
| Have six students sit in a configuration of three rows of two students each. Ask, *How many rows of students?* Give students a chance to answer, **three.** Then say, *one, two, three. There are three rows.* Ask, *How many students are there in a row?* Give students a chance to answer, **two,** and then say, *one, two. Two students are in each row.* Display a 5 × 5 grid. Shade 2 boxes in 3 rows. Say, *This is an **array**. This array shows the number of students.* Emphasize *array*. Create the number sentence and solve it based on the array. | Display a 5 × 5 grid. Shade 3 boxes in 4 rows. Say, *This grid has an array. Each row is 3 boxes, and there are 4 rows.* Emphasize the /z/ sound of rows. Have students count the number of boxes that are shaded. **12** Write on the board: *3 + 3 + 3 + 3 = 12.* Have students get into pairs. Give each pair a 5 × 5 grid and an index card with a repeated addition problem, for example, 2 + 2 + 2. Say, *Use the number sentence to create an array.* Then have students tell the number of rows and columns in their arrays. Ensure they pronounce the /z/ sound when they say rows. | Have students get into groups of three. Give each group a 5 × 5 grid. Say, *Use the grid to create an array. Remember that each row will have the same number of shaded boxes.* Have groups exchange grids, and say, *Create a repeated addition sentence to show how many boxes are in the array.* After students have created and solved their addition sentences, have them describe the arrays they used to another group. |

## Teacher Notes:

NAME _____ DATE _____

# Lesson 5 Guided Writing
## *Repeated Addition with Arrays*

**How do you do repeated addition with arrays?**

Use the exercises below to help you build on answering the Essential Question. Write the correct word or phrase on the lines provided.

**1.** What key words do you see in the question?

array, repeated addition

---

| **Word Bank** | | | |
|---|---|---|---|
| columns | equal groups | number sentence | rows |

**2.** An array is objects displayed in <u>rows</u> and <u>columns</u>.

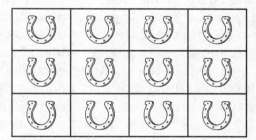

**3.** Arrays can help me count <u>equal</u> <u>groups</u>.

**4.** I can write a <u>number</u> <u>sentence</u> to tell about an array, like this: $4 + 4 + 4 = 12$.

**5.** How do you do repeated addition with arrays?

Arrays can help me put objects in groups. Then I add the groups.

_____

**Teacher Directions:** Read the Building on the Essential Question. Provide descriptions, explanations, or examples of the terms using images or real objects. Read each sentence frame and have students use the lesson and Glossary to write the appropriate terms. Have students read the sentences aloud.

# Lesson 6 Even and Odd Numbers

## English Learner Instructional Strategy

### Language Structure Support: Sentence Frames

Before the lesson, write *odd, even, left over,* and *equal groups* on a chart. Introduce the words, draw a math example, and provide concrete objects to support understanding.

Display the following sentence frames and utilize them during the lesson presentation. **We have _____ cubes,** to describe the value. After evaluating whether the amount is odd or even say: **We have an (even/odd) number of cubes**. Model using the sentence frames and have the students repeat them chorally.

Pair ELs with a native English-speaking student during the Problem Solving exercises and have them echo read the problems.

### English Language Development Leveled Activities

| Emerging Level | Expanding Level | Bridging Level |
|---|---|---|
| **Basic Vocabulary** | **Multiple Meaning Words** | **Public Speaking Norms** |
| On the board, write Even and Odd as the headings in a two-column chart. Use connecting cubes to model the meanings of the words. For example, using 8 connecting cubes, say, *I made 2 equal groups. There are no cubes left. 8 is even.* Write 8 under *Even*. With 7 cubes you would say, *I made 2 equal groups. There is one cube leftover. 7 is odd.* Write 7 under *Odd*. Distribute various numbers of cubes to each student. Have them model the activity. Ask, *even or odd?* Give students a chance to answer, and then write the number on the board. | Discuss multiple meanings of *odd*. Write numbers 1–20 on slips of paper and place them in a container. Have each student select a number. Then they use manipulatives to determine whether the number they selected is odd or even. Say, *If your number is odd, tell me about something odd. If your number is even, tell me about something that is not odd at all.* Provide examples: **The number _____ is odd. I think a green dog is odd. The number _____ is not odd. It is even. Seeing kids playing at a playground is not odd.** | Have students get into pairs. Say, *Write down the number I say as I count you.* Count each student, pointing as you count them, so students know which number they have been assigned. Ask several volunteers, *Is your number odd or even? How do you know?* Have students answer using complete sentences. |

**Teacher Notes:**

NAME _____ DATE _____

# Lesson 6 Word Web

## *Even and Odd Numbers*

Draw a picture in each rectangle that shows the meaning of *even* and *odd*.

| | |
|---|---|
| See students' examples. | See students' examples. |

**even**    **odd**

| | |
|---|---|
| See students' examples. | See students' examples. |

**Teacher Directions:** Provide a description, explanation, or example of the new terms using images or real objects. Have students say each term aloud. Direct students to draw and/or write two math examples to represent each math term. Then encourage students to describe their pictures to a peer.

Grade 2 • **Chapter 2** *Number Patterns*  **21**

# Lesson 7 Sums of Equal Numbers
## English Learner Instructional Strategy

### Collaborative Support: Echo Reading

Review the word *missing* before beginning the lesson. Write the word on the board and discuss/pantomime its meaning. Invite students to share experiences of missing or lost items. Have Expanding and Bridging level students model using the sentence frames: **I lost _____. It was missing.** Encourage Emerging ELs to participate by naming or drawing an item they have lost.

Pair ELs with native English-speakers. Have the English-speaking student read aloud the Problem Solving exercises, then have the EL echo back the problem. Students then work together to solve. Encourage English-speaking peers to explain unfamiliar words to students as needed.

### English Language Development Leveled Activities

| Emerging Level | Expanding Level | Bridging Level |
|---|---|---|
| **Making Connections** Write $4 + 4 = 8$ on the board. Use connecting cubes to model solving the problem. Hold both sets of 4 cubes side by side. *Say, 4 cubes and 4 cubes. There are two* equal groups. Emphasize *equal groups.* Connect the cubes. Say, *I put 4 cubes and 4 cubes together to make 8. Two groups of 4 cubes equals 8 cubes.* Repeat with other examples. | **Share What You Know** Write *equal* on the board. Write even numbers under 20 on pairs of index cards. Distribute a card to each student. Say, *Find your equal.* Emphasize the /kw/ sound in equal. Give students time to find their partner with the same number. Then say, *Work together to find the sum of your two equal addends.* Give students several minutes to complete the task. Have each pair describe their addends using the sentence frames: _____ **and** _____ **are equal addends. Their sum is** _____. Be sure students say the /kw/ sound in *equal*. | **Academic Word Knowledge** Write the numbers 1 through 20 on slips of paper and place them in a container. Have each student draw a number. Ask, *If your number is a sum, can it have equal addends? How do you know?* Give students a few minutes to think about their answers. Then have volunteers answer in complete sentences. For example, **My number is 15. It will not have equal addends because it is an odd number.** |

### Teacher Notes:

NAME _____ DATE _____

# Lesson 7 Note Taking
## Sums of Equal Numbers

Read the question. Write words you need help with.
Use your lesson to write your Cornell notes. Write or
draw math examples to explain your thinking. Share
your examples with a classmate.

| **Building on the Essential Question** | **Notes:** |
|---|---|
| How can I find equal addends of even numbers? | Numbers that end with 0, 2, 4, 6, and 8 are __even__ __numbers__.<br><br>I can break an even number of objects into 2 __equal__ __groups__ like this:<br><br>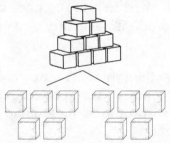<br><br>An even number can be written as the __sum__ of two __equal__ __addends__.<br><br>$$10 = \underline{\phantom{5}5\phantom{5}} + \underline{\phantom{5}5\phantom{5}}$$ |
| **Words I need help with:**<br>See students' words. | |

**My Math Examples:**
See students' examples.

**Teacher Directions:** Read the Building on the Essential Question and have students list words/phrases they need assistance with. Provide descriptions, explanations, or examples of the terms using images or real objects. Read each sentence frame and have students write the appropriate terms. Have students read their notes aloud. Direct students to create an example representing the question. Then encourage students to describe their picture to a peer.

# Chapter 3 Add Two-Digit Numbers

## What's the Math in This Chapter?

### Mathematical Practice 8: Look for and express regularity in repeated reasoning

Write the addition problem 34 + 62 horizontally on the board. Ask students to share strategies to solve this problem. Rewrite the problem vertically in a place-value chart.

Ask, *Why does writing the math problem vertically make it easier to add than when it is shown horizontally?* **We can line up place value.** Model solving the problem and discuss with students that adding larger numbers is similar to addition problems that they have already done. Point out to students that they will follow a series of steps to add two-digit numbers, i.e., begin by adding the ones column and then add the tens column.

Ask, *How is adding the numbers in the ones and tens places similar?* The discussion should focus on repeated calculation. Highlight that these steps are like rules that need to be followed when solving multi-digit addition problems. Students will need to add the ones digits, regroup if necessary, add the tens digits, regroup if necessary and continue on with other place values.

Display a chart with Mathematical Practice 8. Restate Mathematical Practice 8 and have students assist in rewriting it as an "I can" statement, for example: **I can solve problems by looking for rules and patterns.** Post the new "I can" statement.

## Inquiry of the Essential Question:

### How can I add two-digit numbers?

Inquiry Activity Target: **Students come to a conclusion that they can apply what they already know about addition to add larger numbers.**

As an introduction to the chapter, present the Essential Question to students. The inquiry graphic organizer will offer opportunities for students to observe, make inferences, and apply prior knowledge of addition processes representing the Essential Question. As they investigate, encourage students to draw, write, and collaborate with peers to demonstrate their observations and thinking. Then have students present additional questions they may have to a peer to extend discussions.

Regroup students and restate Mathematical Practice 8 and the Essential Question. Pose questions to reflect on what has been learned to guide students in making connections between the Mathematical Practice and the Essential Question.

NAME _____ DATE _____

# Chapter 3 Add Two-Digit Numbers
## *Inquiry of the Essential Question:*

**How can I add two-digit numbers?**

Read the Essential Question. Describe your observations
(I see...), inferences (I think...), and prior knowledge (I know...)
of each math example. Write any questions you have.

$$
\begin{array}{r}
1 \\
21 \\
13 \\
37 \\
+\ 25 \\
\hline
96
\end{array}
$$

$3 + 7 = 10$
$10 + 1 = 11$
$11 + 5 = 16$

I see ...

I think ...

I know ...

---

$37 + 45$

$32 + 5 + 45$

$32 + 50 \quad = 82$

I see ...

I think ...

I know ...

---

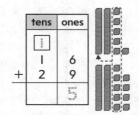 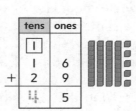

I see ...

I think ...

I know ...

---

Questions I have...

_____

_____

 **Teacher Directions:** Read the Essential Question for students. Have students echo read. Direct students to describe their observations, inferences, and prior knowledge of each math example. Encourage students to write or draw additional questions they may have. Then have students share their ideas/questions with a peer.

Grade 2 • **Chapter 3** *Add Two-Digit Numbers* **23**

# Lesson 1 Take Apart Tens to Add

## English Learner Instructional Strategy

### Vocabulary Support: Build Background Knowledge

Write *take apart* in the center of a word web on chart paper. Encourage students to share something they have taken apart. Write their experiences on the word web to help build background knowledge. As you add the experiences to the web invite students to draw illustrations as well. Invite emerging students to draw an experience with *take apart* if they are not ready to share verbally. Pair same native language emerging level students with students having a higher level of English proficiency. Have the pairs work together to solve the problems on the See and Show page. Invite them to report their answers with the sentence frames: **I took apart** ____ **to get** ____ **+** ____. **I added** ____ **and** ____ **to get the sum** ____.

### English Language Development Leveled Activities

| Emerging Level | Expanding Level | Bridging Level |
| --- | --- | --- |
| **Word Knowledge**<br><br>Show 12 connected cubes. Say, *I can **take apart** all the cubes.* Take apart the cubes so you have twelve unconnected cubes. Reconnect the cubes, and say, *I can **take apart** the cubes so I have 10 and 2.* Take apart the cubes so you have a set of 10 and a set of 2. Show the cubes to students as you say, *10 and 2.* Give each student a set 17 connected cubes. Say, *Take apart the cubes so you have 10 and 7.* Give students a moment to complete the task, and then model. Repeat the activity with other amounts of connected cubes. | **Sentence Frames**<br><br>Post these sentence frames: **Take apart** ____ **as** ____ **and** ____. **Then make** ____ **plus** ____ **into** ____. **Finally add** ____ **plus** ____ **to get** ____. Write 19 + 8 on the board. Model solving the problem by taking apart tens to add while modeling the use of the sentence frames. Then write 34 + 16 = ____. Have students use the sentence frames to guide you in solving the new problem. Repeat the activity with 66 + 24 = ____ and have students use the frames to solve the problem on their own. | **Exploring Language Structure**<br><br>Write 23 + 18 on the board. Say, *I will **take** apart 23 to make this problem easier to solve.* Emphasize *take* apart. Then model solving the problem by taking apart tens to add. After you finish, say, *I **took** apart 23. I had 21 and 2. I added 2 and 18 to get 20. Then I added 20 and 21 to get 41.* Emphasize *took*. Write 37 + 14 on the board. Have students work in pairs to solve the problem by taking apart tens to add. Then have a volunteer describe how they solved the problem. Be sure they correctly use the irregular past tense. |

### Teacher Notes:

NAME _____ DATE _____

# Lesson 1 Word Identification
## *Take Apart Tens to Add*

Match each term to the example.

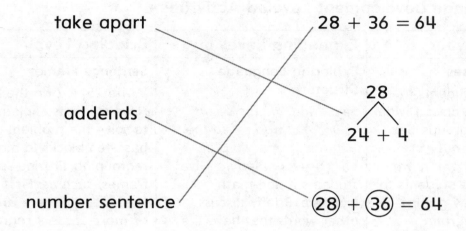

take apart             28 + 36 = 64

                                28

addends                     24 + 4

number sentence      (28) + (36) = 64

---

Write the correct term from above for each sentence on the blank lines.

**1.** 32 + 39 = 71 is a __number__ __sentence__ .

**2.** I can __take__ __apart__ addends to make numbers that are easier to add.

**3.** When I add two __addends__ , I get a sum.

**Teacher Directions:** Review the terms using images or real objects. Have students say each term and then draw a line to match the term to its example. Direct students to say each term and then write the corresponding meanings in the sentences. Encourage students to read the sentences to a peer.

# Lesson 2 Regroup Ones as Tens
## English Learner Instructional Strategy

### Vocabulary Support: Model Academic Vocabulary

Write *regroup, ones,* and *tens* on the board. Add *regroup* and its Spanish cognate, *reagrupar,* to your cognate chart. Review the words *ones* and *tens* using a place-value chart. Introduce *regroup,* and model a math example using base-ten blocks to support understanding. Assist non-Spanish speaking ELs with an appropriate translation tool.

### English Language Development Leveled Activities

| Emerging Level | Expanding Level | Bridging Level |
|---|---|---|
| **Choral Responses** Invite several students to the front of the class. Divide them into two groups. Say, *Here are two groups of students. Now I will regroup them.* Direct the students to get into two new groups. Say, *When I regroup, I make new groups.* Have students return to their seats. Write 14 + 7 on the board. Use base-ten blocks (rods and units) to show a group of 14 and a group of 7. Before the regrouping step, ask, *Should I regroup?* Give students a moment to answer **yes** chorally, then say, *Yes, I will regroup.* Finish solving the problem. Repeat with 36 + 6. | **Exploring Language Structure** Say, *"Re-" means again. When you add re- to the beginning of a word, it changes its meaning.* Give students a non-math example. Briefly discuss other words that have the re- prefix. Make sure students understand that not all words beginning with re- use the prefix. Say, *I will write a problem on the board.* Write 26 + 8 on the board, but then erase it. Say, *I will rewrite the problem. I will write it again.* Emphasize re-. Then model solving the problem with base-ten blocks. Prompt students to tell you when it is time to regroup. | **Sentence Frames** Write 25 + 7 on the board. Have students explain how to solve the problem using base-ten blocks to model regrouping. Provide sentence frames, such as: **First I add the ____. If there are ten or more ____, I regroup ten ____ as one ten. I now have ____ tens and ____ ones. The sum is ____.** Repeat the activity with another addition problem that requires regrouping. |

### Multicultural Teacher Tip

During the lesson, you may experience EL students who appear to listen closely to your instructions and exhibit verbal and/or nonverbal confirmation that they understand the concepts. However, it becomes clear during the See and Show or On My Own parts of the lesson that the students did not actually understand. This may be due to a student coming from a culture in which the teacher is regarded as a strong, perhaps even intimidating, authority figure. They may be reluctant to ask questions, considering it impolite to do so and an implication that the teacher is failing.

NAME _____ DATE _____

# Lesson 2 Vocabulary Definition Map
## *Regroup Ones as Tens*

Use the definition map to write what the math word means and tell what the word is like. Write or draw a math example. Share your examples with a classmate.

**My Math Word:**

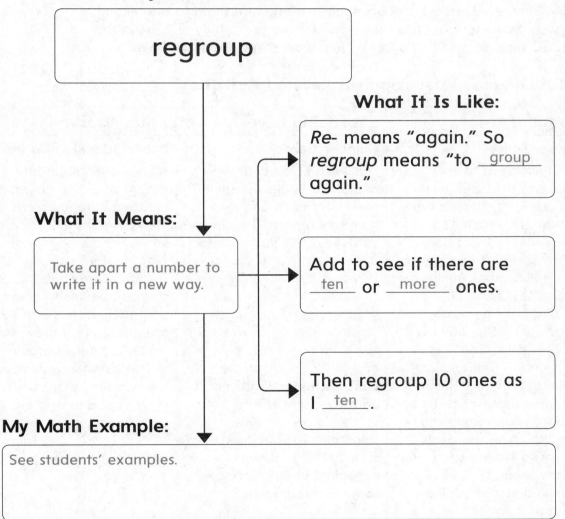

regroup

**What It Is Like:**

*Re-* means "again." So *regroup* means "to _group_ again."

**What It Means:**

Take apart a number to write it in a new way.

Add to see if there are _ten_ or _more_ ones.

Then regroup 10 ones as 1 _ten_.

**My Math Example:**

See students' examples.

**Teacher Directions:** Provide a description, explanation, or example of the new term using images or real objects. Have students use the lesson or Glossary to define the math term. Direct students to list characteristics, and draw a picture representing their math term. Then encourage students to describe their picture to a peer.

Grade 2 • **Chapter 3** *Add Two-Digit Numbers* **25**

# Lesson 3 Add to a Two-Digit Number

## English Learner Instructional Strategy

### Graphic Support: Word Webs

For Developing Vocabulary, write the term *add* in a word web. Say, *I can add lemon to my water before I drink it.* Have students give examples of things they can add to something else using the sentence frame: **I can add _____ to _____.** Add student ideas to the web.

Review signal words for addition with students by creating another word web for *add*. Have students work in small groups to brainstorm words, symbols, and terms that mean they need to add, to complete a math problem. Invite volunteers to share their lists and add them to the second word web. Make sure the final web includes the terms they will see in the exercises, such as: *in all, altogether, join, how many... now, and plus.*

### English Language Development Leveled Activities

| Emerging Level | Expanding Level | Bridging Level |
|---|---|---|
| **Number Sense** | **Number Game** | **Public Speaking Norms** |
| Use index cards to create a set of one-digit and two-digit numbers. Show a one-digit number, such as 7, and say, *seven. This is a **one-digit** number.* Then show a two-digit number, such as 35, and say, *thirty-five. This is a **two-digit** number.* Randomly display numbers one at a time, and ask, *Is this a one-digit number or a two-digit number?* Have students answer chorally or with a gesture. Once you have shown all the cards and asked students to identify them, choose two cards and model adding a two-digit number and a one-digit number. | Pair each EL student with a mentor or English-speaking peer. Give each pair 3 number cubes and an index card. Say, *Roll your number cubes. Use two of the numbers you roll to create a two-digit number. Write a vertical addition problem on the index card that adds the third number to the two-digit number.* Have pairs exchange index cards, and say, *Solve the problem you were given.* Have volunteers explain how they solved their problems, including whether or not they used regrouping. | Have students get into groups of 3. Tell groups to create a story problem. Say, *Your story problem will be solved by adding a one-digit number to a two-digit number.* Give students several minutes to come up with their stories and solve the addition problem. Have students refer to problems in the lesson for examples. Have groups share their story problems and answers with the class. |

### Teacher Notes:

NAME _____ DATE _____

# Lesson 3 Concept Web

## *Add to a Two-Digit Number*

Use the concept web to show what you know about regrouping. Circle *yes* or *no*.

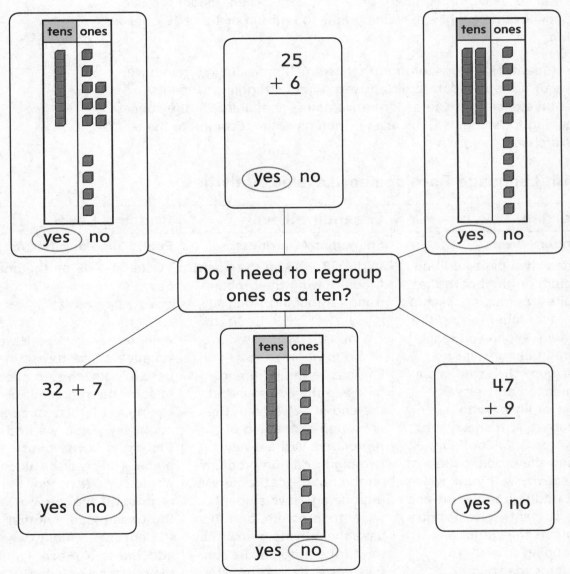

**Teacher Directions:** Provide a description, explanation, or example of the term. Practice the sentence frame **[Yes/No], I [need/don't need]** to regroup ten ones as I ten. After students have completed the web, have them orally compare answers with a partner.

# Lesson 4 Add Two-Digit Numbers

## English Learner Instructional Strategy

### Sensory Support: Manipulatives

Before the lesson, have a native language peer or mentor work with emerging/expanding students to introduce Work Mat 6 used in the Add Two-Digit Numbers lesson. Say in English or native language, *This is a tool that makes it easier to add 2 two-digit numbers. We can use the work mat to keep the ones and tens in their own spaces. We can also use the work mat to show addition with base-ten blocks.* Then, model the See and Show example while explaining how to regroup or trade 10 ones for one ten.

Have students first work through Exercises 6–11 individually. Then have them pair up to compare their answers. Direct struggling students to model an exercise for you as formative assessment. Reteach the concept of regrouping for students if necessary, then have them complete the rest of the exercises.

### English Language Development Leveled Activities

| Emerging Level | Expanding Level | Bridging Level |
|---|---|---|
| **Number Sense** | **Phonemic Awareness** | **Public Speaking Norms** |
| Write a two-digit addition problem on the board that requires regrouping, such as 27 + 14. Model solving the problem. When you reach the regrouping step, ask, *Do I regroup?* Give students a moment to answer **yes**, either orally or with a gesture, and then say, *Yes, I will regroup.* Continue solving the problem. Repeat the activity with new two-digit addition problems and have students solve. Provide problems that require regrouping as well as problems that do not. | Write 38 + 15 on the board. Model solving the problem using regrouping. Then say, *My answer is 53. I added, regrouped, and then added again to get my answer.* Emphasize the difference between the /d/ sound at the end of *added* and the /t/ sound at the end of *regrouped.* Write a new two-digit addition problem on the board that requires regrouping. Have students solve the problem, and then have them explain how they solved the problem. Be sure they correctly say the /d/ and /t/ sounds to indicate past tense. | Write 34 + 18 on the board. Say, *Please tell me the steps I need to follow to solve this problem. Use complete sentences.* Have volunteers take turns describing each step needed to solve the problem. For example, the first student might say, *First,* **we add the digits in the ones place.** The second student would say, *Next,* **we regrouped the number in the ones place.** Another student says, *Finally,* **we add the tens place.** Prompt the correct words that indicate order as necessary. |

### Teacher Notes:

NAME _____ DATE _____

# Lesson 4 Guided Writing

## *Add Two-Digit Numbers*

### How do you add two-digit numbers?

Use the exercises below to help you build on answering the Essential Question. Write the correct word from the word bank on the lines provided.

I.  **What key words do you see in the question?**
    two-digit numbers
    _____

| Word Bank | | | |
|---|---|---|---|
| ones | regroup | tens | two-digit number |

2.  A digit is a symbol used to write numbers. The number 37 is a __two - digit__ __number__.

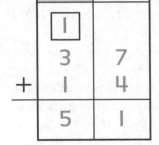

Find 37 + 14.

|       | tens | ones |
|-------|------|------|
|       | 1    |      |
|       | 3    | 7    |
| +     | 1    | 4    |
|       | 5    | 1    |

3.  Add the __ones__ first. If there are more than 10 ones, __regroup__ 10 ones as 1 ten.

4.  Then add the __tens__.

5.  How do you add two-digit numbers?

    First add the ones. Regroup. Add the tens.
    _____

    _____

**Teacher Directions:** Read the Building on the Essential Question. Provide descriptions, explanations, or examples of the terms using images or real objects. Read each sentence frame and have students use the lesson and Glossary to write the appropriate terms. Have students read the sentences aloud.

# Lesson 5 Rewrite Two-Digit Addition
## English Learner Instructional Strategy

### Vocabulary Support: Build Background Knowledge

Before the lesson, work with ELs to understand the potentially unfamiliar English terms in this lesson. Use photographs and pictures to build students' understanding of the words: *baseball, hot dogs, hamburgers, game, team, children's hospital, scored,* and *runs.* Provide a translation tool for students who still need clarification.

Discuss *raising money.* Invite students to share other ways to raise money using the sentence frame: **I can raise money by _____.**

### English Language Development Leveled Activities

| Emerging Level | Expanding Level | Bridging Level |
|---|---|---|
| **Building Word Knowledge** Write 32 + 16 horizontally on the board. Then say, *I am going to write this problem again. I am going to rewrite this problem.* Emphasize the prefix *re-.* Rewrite the problem horizontally again. Then ask, *Did I rewrite the problem in the same way or a different way? Same or different?* Give students a moment to answer, **same,** then say, *I rewrote the problem in the same way.* Say, *I will rewrite the problem a different way.* Write 32 + 16 vertically on the board. Repeat the questions above, encouraging students to answer **different**. | **Number Game** Using index cards, create sets of two-digit addition problems. One card will show the problem written horizontally and one will show the problem written vertically. Randomly distribute the cards to students. Say, *Look at the problem on your card. Find the student with the same problem written a different way.* Give students time to locate their partners, and then have them solve the problem. | **Listen and Write** Invite two students to the board. Give one student an index card with a two-digit addition problem written horizontally. Tell the student to explain each step needed to rewrite the problem vertically and to solve it. The second student will follow the steps to rewrite the problem on the board and solve it. Invite a new pair of students to the board and repeat the activity with a new two-digit addition problem. |

### Teacher Notes:

NAME _____ DATE _____

# Lesson 5 Note Taking
## *Rewrite Two-Digit Addition*

Read the question. Write words you need help with.
Use your lesson to write your Cornell notes. Write or
draw math examples to explain your thinking.

| **Building on the Essential Question** | **Notes:** |
|---|---|
| How can I rewrite two-digit addition? | *Re-* means "again," so *rewrite* means "<u>write</u> <u>again</u>." <br><br> I can <u>rewrite</u> a problem to add. <br><br> $\boxed{\text{Find } 34 + 48.}$ <br><br> Write one addend <u>below</u> the other addend. <br><br> Line up the <u>tens</u> and <u>ones</u>. <br><br> 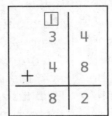 <br><br> Add. <br><br> If there are more than 10 ones, I should <u>regroup</u> as 1 <u>ten</u>. |
| **Words I need help with:** <br> See students' words. | |

**My Math Examples:**
See students' examples.

 **Teacher Directions:** Read the Building on the Essential Question and have students list words/phrases they need assistance with. Provide descriptions, explanations, or examples of the terms using images or real objects. Read each sentence frame and have students write the appropriate terms. Have students read their notes aloud. Direct students to create an example representing the question, and then describe it to a peer.

# Lesson 6 Add Three or Four Two-Digit Numbers

## English Learner Instructional Strategy

### Vocabulary Support: Communication Guide

Post the following communication guide for students to utilize during the lesson.

The numbers in the ones place are...
I can first make a ten with _____ and _____.
I can use a doubles fact and add _____ and _____ first.
The numbers in the tens place are...
I will regroup because...
The sum is _____.

Invite students at lower levels of proficiency to write or point at their answers or use manipulatives to show their thinking.

### English Language Development Leveled Activities

| Emerging Level | Expanding Level | Bridging Level |
|---|---|---|
| **Cooperative Learning** | **Phonemic Awareness** | **Show What You Know** |
| On the board, create a two-column chart labeled Tens and Ones. Divide the students into two groups; Tens and Ones. Give each group a (0–5) number cube, have them roll it and write the number rolled in their assigned column. Guide the students as needed. Repeat until three, two-digit numbers have been generated. Use the numbers to model solving an addition problem. Prompt student responses with questions, such as, *Can I make a ten?*, *Should I add these doubles facts first?*, *Do I regroup?*, and *What do I do now?* | Write 34 + 28 + 19 on the board. As you model solving the problem, say, *First I add the ones column. Then I regroup the ones. Then I add the tens column.* Emphasize the /z/ sound at the end of the plurals. Write a new addition problem with three or four two-digit numbers. Invite volunteers to the board to solve the problem. Encourage them to explain each step as they solve the problem. Be sure they are correctly saying the plural /z/ sound for *ones* and *tens*. | Write 14 + 35 + 19 + 23 on the board. Have students get into pairs, and then say, *Use this addition number sentences to create a word problem. Then solve the addition number sentence.* Give pairs time to come up with their word problem and solve the addition number sentence. Then have one student from each pair tell their word problem and explain how they solved the addition number sentence. |

**Teacher Notes:**

NAME _____ DATE _____

# Lesson 6 Note Taking

## *Add Three and Four Two-Digit Numbers*

Read the question. Write words you need help with.
Use your lesson to write your Cornell notes. Write or
draw math examples to explain your thinking.

| **Building on the Essential Question** | **Notes:** |
|---|---|
| How can I add three and four two-digit numbers? | ___Line___ ___up___ the ones and tens.  $\begin{array}{r}\boxed{1}\phantom{0} \\ 3\,|\,2 \\ 2\,|\,8 \\ +1\,|\,8 \\ \hline 7\,|\,8 \end{array}$ |
| | If there are more than 10 ___ones___, I should ___regroup___ as 1 ___ten___. |
| | When I add, I should look for a ___fact___ I know. |
| **Words I need help with:** | I can make a ___10___ like this: |
| See students' words. | $\begin{array}{r}3\,②\\2\,⑧\\+1\ 8\end{array}\Big\rangle\ 2 + 8 = 10$ |
| | I can also look for ___doubles___ : |
| | $\begin{array}{r}3\ 2\\2\,⑧\\+1\,⑧\end{array}\Big\rangle\ 8 + 8 = 16$ |

**My Math Examples:**

See students' examples.

**Teacher Directions:** Read the Building on the Essential Question and have students list words/phrases they need assistance with. Provide descriptions, explanations, or examples of the terms using images or real objects. Read each sentence frame and have students write the appropriate terms. Have students read their notes aloud. Direct students to create an example representing the question, and then describe it to a peer.

**Grade 2 · Chapter 3** *Add Two-Digit Numbers* **29**

# Lesson 7 Problem Solving Strategy: Make a Model

## English Learner Instructional Strategy

### Collaborative Support: Partners Work

Before the lesson, review any potentially unfamiliar vocabulary terms in the Problem Solving exercises, such as: *walk-a-thon, basketball, beach, shells, ran, blocks, action figures, points, pairs of earrings, muffins, glass bottles,* and *recycle.* Use photos, and realia to support understanding.

Have students work in multilingual pairs. Assign Exercises 1–6. Direct students to first read aloud the problem either simultaneously or using echo reading. Then have them discuss the different strategies they could use to solve the problem.

### English Language Development Leveled Activities

| Emerging Level | Expanding Level | Bridging Level |
|---|---|---|
| **Background Knowledge** | **Building Oral Language** | **Graphic Organizer** |
| Write: Understand, Plan, Solve, and Check. Write then read aloud: *There are 18 apples in a basket. 27 apples are added to the basket. How many apples are in all?* Point to Understand on the board. Then ask, *Do we know how many apples in all?* Give students time to respond, **no** and then say, *I will circle what we need to find.* Circle the last sentence. Point to Plan, and say, *Now we need a plan. We can make a model to solve.* Write *Make a Model* next to Plan on the board. Continue in this manner as you model solving the problem. | Draw four large boxes on the board, and label them Understand, Plan, Solve, and Check. In the Understand box, write then read aloud: *Ruby has 12 books checked out from the library. She checks out 24 more. How many books does Ruby have checked out?* Guide students as they help you complete the graphic organizer. Provide sentence frames, such as: **We know ____ and ____. You should underline ____. We need to find out ____. You should circle ____.** Guide students to choose making a model as their plan for solving the problem. | Draw a large problem solving graphic organizer. In the Understand box, write then read aloud: *At the first stop, 13 children got on the bus. 9 children got on at the second stop. 18 children got on the bus at the third stop. How many children are on the bus?* Ask students to identify what they know. Have them underline the first three sentences of the problem. Ask, *Do we know how many students are on the bus?* Give students a chance to answer **no**. Have each student make a model to solve the problem. Share all model ideas with the group. |

### Teacher Notes:

NAME _____ DATE _____

# Lesson 7 Problem Solving
## STRATEGY: Make a Model

<u>Underline</u> what you know. (Circle) what you need
to find. Make a model to solve the problems.

**I. Karenna** ran **I5** blocks Tuesday.

**She** (Karenna) ran **20** blocks on
Wednesday.

How many blocks did she run in all?

street blocks

$$
\begin{array}{r}
\phantom{0} \\
+ \phantom{00} \\
\hline
\end{array}
$$

She ran __35__ blocks in all.

---

**2. Bradley got 33** action figures for his
birthday.

**He** (Bradley) **had I2** already.

How many will he **have** if he
**buys** I5 more?

action
figure

$$
\begin{array}{r}
\phantom{0} \\
+ \phantom{00} \\
\hline
\end{array}
$$

He will have __60__ action figures.

**Teacher Directions:** Provide a description, explanation, or example of the boldface terms
and nouns using images or real objects. Read each sentence and have students echo read.
Encourage students to make a model using manipulatives and then write their answers in
each restated question. Have students read each answer sentence aloud.

# Chapter 4 Subtract Two-Digit Numbers

## What's the Math in This Chapter?

### Mathematical Practice 8: Look for and express regularity in repeated reasoning

Tell students that you started school with 36 pencils and gave 14 away to your friends. Ask, *What math problem would I need to solve to find out how many pencils I have left?* **36 — 14** Write the subtraction problem 36 — 14 on the board vertically in a place-value chart. Ask students to share strategies to solve this problem.

Model solving the problem and discuss with students that subtracting larger numbers is similar to other math problems they have solved. Point out to students that they will follow a series of steps to solve this problem, just like they did when they added multi-digit numbers, i.e., begin by subtracting the ones column and then subtracting the tens column.

Ask, *How is subtracting the numbers in the ones and tens places similar?* The discussion should focus on repeated calculation. Highlight that these series of steps need to be followed when solving multi-digit subtraction problems. Students will need to subtract the ones digits, regroup if necessary, subtract the tens digits, regroup if necessary and continue on with other place values.

Display a chart with Mathematical Practice 8. Restate Mathematical Practice 8 and have students assist in rewriting it as an "I can" statement, for example: **I can follow a series of steps to solve a problem.** Post the new "I can" statement.

## Inquiry of the Essential Question:

### How can I subtract two-digit numbers?

Inquiry Activity Target: **Students come to a conclusion that they can apply what they already know about subtraction to subtract larger numbers.**

As an introduction to the chapter, present the Essential Question to students. The inquiry graphic organizer will offer opportunities for students to observe, make inferences, and apply prior knowledge of the subtraction process representing the Essential Question. As they investigate, encourage students to draw, write, and collaborate with peers to demonstrate their observations and thinking. Then have students present additional questions they may have to a peer to extend discussions.

Regroup students and restate Mathematical Practice 8 and the Essential Question. Pose questions to reflect on what has been learned to guide students in making connections between the Mathematical Practice and the Essential Question.

NAME _____ DATE _____

# Chapter 4 Subtract Two-Digit Numbers
## *Inquiry of the Essential Question:*

**How can I subtract two-digit numbers?**

Read the Essential Question. Describe your observations (I see...), inferences (I think...), and prior knowledge (I know...) of each math example. Write any questions you have.

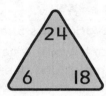

| tens | ones |
|------|------|
| 4    | 13   |
| 5    | 3    |
| − 1  | 7    |
| 3    | 6    |

I see ...

I think ...

I know ...

---

24
6    18

6 + 18 = 24      24 − 6 = 18

18 + 6 = 24      24 − 18 = 6

I see ...

I think ...

I know ...

---

**Subtract**

75   Add these
− 38  numbers
      to check.
37

**Check by adding**

37   If this is the number
     you subtracted from,
+ 38  your answer is
75 ←correct.

I see ...

I think ...

I know ...

---

Questions I have...

_____

_____

**Teacher Directions:** Read the Essential Question for students. Have students echo read. Direct students to describe their observations, inferences, and prior knowledge of each math example. Encourage students to write or draw additional questions they may have. Then have students share their ideas/questions with a peer.

**Grade 2 · Chapter 4** *Subtract Two-Digit Numbers*   **31**

# Lesson 1 Two-Digit Fact Families
## English Learner Instructional Strategy

### Vocabulary Support: Activate Prior Knowledge

Use the Math Word Wall to review the terms: *addition, addend, sum, subtraction, difference,* and *fact family.* Organize students into groups of 3–5 and assign a term to each group. Have students look up the term in the Glossary. Provide students with chart paper to write a definition, and draw a visual math example with the term labeled. Then, invite a volunteer from each group to share their definitions using the sentence frame: **Our math word is ____. It means ____.** Post the charts in the classroom, and have students record the terms and definitions in a journal for future reference.

For Problem Solving Exercises 9 and 10, pair emerging level students with same native language bridging level students. Encourage use of native language to clarify meaning.

### English Language Development Leveled Activities

| Emerging Level | Expanding Level | Bridging Level |
|---|---|---|
| **Making Connections** | **Look, Listen, and Identify** | **Partners Work/Pairs Share** |
| Ask, *Who is in a family? Is a mother in a family?* Give students time to answer **yes**, then say, *Yes, a mother is in a family. Is an uncle in a family?* Continue with a few more examples. Say, *The people in a family are* **related**. Write an example of a two-digit fact family, such as 17, 8, and 9. Write the four equations that show these numbers are a fact family. Ask, *Is this a fact family?* Give students time to answer, **yes**. Say, *Yes, this is a fact family. The numbers in a fact family are related.* Continue with examples and non-examples of fact families. Have students determine whether the number sentences are a fact family or not. | On the board, write: *Fact Families and Not Fact Families.* Then use self-stick notes to write examples and non-examples of two-digit fact families. Pair students, and randomly distribute one sticky note to each pair. Say, *Use the numbers on your sticky notes to write addition and subtraction sentences. Do you have a fact family on your card or not?* Give students time to answer. Have one student from each pair come to the board and place their sticky note below the correct category. Then briefly point out that *-ies* is added to family to form the plural. | Have students work in pairs. Write examples and non-examples of fact families on index cards, and then randomly distribute one index card to each pair. Ask, *Do you have a fact family? How do you know?* Give students several minutes to solve equations to find the answer. Then have pairs share with another pair of students to identify whether they have a fact family or not, and how they know, using the sentence frame: **This ____ (is/is not) a fact family because ____.** |

**Teacher Notes:**

NAME _____ DATE _____

# Lesson 1 Concept Web
## *Two-Digit Fact Families*

Use the concept web to show what you know about fact families. Circle *yes* or *no*.

yes     no

12 + 13 = 25

13 + 12 = 25

13 − 12 = 1

yes     (no)

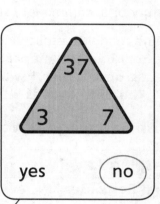

yes     (no)

## Is it a fact family?

27 + 13 = 40

13 + 27 = 40

40 − 27 = 13

40 − 13 = 27

(yes)     no

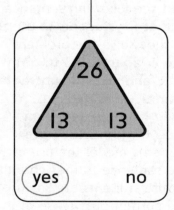

(yes)     no

12 + 12 = 24

24 − 12 = 12

(yes)     no

**Teacher Directions:** Provide a description, explanation, or example of the term. Practice the sentence frame **[Yes/No], it [is/is not] a fact family.** After students have completed the web, have them orally compare answers with a partner.

# Lesson 2 Take Apart Tens to Subtract
## English Learner Instructional Strategy

### Sensory Support: Manipulatives

Before the lesson, write the review terms: *ones, tens*, and *take apart* on a chart. Discuss the terms, write math examples, then model with a place-value chart and base-ten blocks to support understanding. Have students write the terms and draw a visual example in their math journals. Next, have students write definitions and sentences using each term in context. Provide sentence examples if needed. Have non-Spanish speaking ELs, refer to the Multilingual eGlossary or accommodate students with a translation tool.

Post and model the following sentence frames to assist students during the lesson: **I will take apart _____ to make a ten. I can make _____ tens and _____ ones. Then I will subtract the tens. Then I will subtract the ones. The difference is _____. So, _____ minus equals _____.**

### English Language Development Leveled Activities

| Emerging Level | Expanding Level | Bridging Level |
|---|---|---|
| **Choral Responses** | **Support Language Structure** | **Communications Guides** |
| Write the symbol (−) alongside the word *subtract*. Point to the symbol, and say, *subtract*. Repeat, saying the word slowly and running your finger below the word on the board. Write several subtraction and addition sentences on the board. Point to them one at a time as you ask, *Do you subtract?* Give students time to say or gesture **yes** or **no**, then say either, *Yes, we will subtract.* or *No, we will not subtract. We will add.* Invite a volunteer to solve each number sentence before moving to the next one. | Write *64 − 32 = _____*. Say, *We can take apart 32 to subtract.* Model solving the problem. Then say, *I took apart 32 to get 30 and 2. I subtracted 30 from 64 to get 34. Then I subtracted 2 from 34 to get 32.* Write 47 − 16. Post the following sentence frames for students to utilize as they solve the problem: **I took apart _____ to get _____ and _____. I subtracted from _____ to get _____. Then I subtracted from _____ to get _____.** Have students model the use of the sentence frames chorally. Ensure students are saying the correct past tense of *take*. | On the board, write: *First I _____. Then I _____. Last I _____.* Tell students that sequence words help show the order that things happen. Write a two-digit subtraction problem on the board that can be solved by taking apart tens, for example, 49 − 13. Ask students to solve the problem. Then have a volunteer use the sentence frames on the board to explain how they solved the problem. Repeat the activity with a new subtraction sentence and a new volunteer. |

### Teacher Notes:

NAME _____ DATE _____

## Lesson 2 Vocabulary Sentence Frames
*Take Apart Tens to Subtract*

The math words in the box are for the sentences below. Write the words that fit in each sentence on the blank lines.

| difference | subtract | take apart |
|---|---|---|

**1.** You can __take__ __apart__ numbers to subtract.

**2.** The answer to a subtraction problem is the __difference__.

$$44 - 13 = \bigcirc{31}$$

**3.** To __subtract__ means to take away, take apart, separate, or find the difference between two sets.

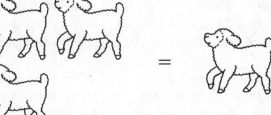

**Teacher Directions:** Provide a description, explanation, or example of the each term using images or real objects. Read each sentence frame and have students echo read. Direct students to write the correct terms in each blank. Then encourage students to read each sentence to a peer.

**Grade 2 • Chapter 4** *Subtract Two-Digit Numbers* **33**

# Lesson 3 Regroup a Ten as Ones
## *English Learner Instructional Strategy*

### Vocabulary Support: Cognates

Write *regroup* on the board, and add it and its Spanish cognate, *reagrupar*, to your cognate chart. Introduce the word, and model the following activity to support understanding. Assist non-Spanish speaking ELs to use the Multilingual eGlossary or accommodate students with an appropriate translation tool.

Before starting the Explore and Explain page, give each student 23 straws. Have the students bundle tens using rubber bands. Students should discover they all have a group of 2 tens and a group of 3 ones. Ask, *How many straws are there altogether?* **23** Have students take apart a group of tens and place it with the group of ones. Discuss that everyone now has a group of 1 ten and a group of 13 ones. Say, *We regrouped 1 ten. How many straws do we still have altogether?* **23** Discuss regrouping.

### English Language Development Leveled Activities

| Emerging Level | Expanding Level | Bridging Level |
|---|---|---|
| **Background Knowledge** | **Building Oral Language** | **Group Activity** |
| Write *compare*. Say, *I am going to compare.* Invite two students to the front of the class. Model comparing by describing differences in the colors of their clothes. Write 5 and 9 on the board. Say, *We will compare the numbers. Which number is more?* Give students time to answer **9,** then say, *9 is more than 5. 5 is less than 9. Can we subtract 9 from 5?* **no** Then say, *No, we cannot subtract 9 from 5 because 9 is more than 5.* Model an example with manipulatives. Repeat with a new pair of one-digit numbers. | On index cards, write two-digit subtraction sentences. Some of the subtraction sentences should require regrouping, and some should not. Assign students to work in pairs, and randomly distribute an index card to each pair. Give pairs a few minutes to solve the subtraction sentences. Ask, *Did you need to regroup? Why or why not?* Have one student from each pair explain their answer. Provide sentence frames, such as: **We had to regroup because _____.** or **We did not have to regroup because _____.** | Have students work in groups of three, and give each group a (0–5) number cube. Have each student in the group roll the number cube once and record the number. Say, *Use two of your group's numbers to create a two-digit number. Subtract the third number you rolled from the two-digit number.* Give students time to solve the problems. Ask, *Did you need to regroup?* Have a student from each group explain whether regrouping was needed. |

### Teacher Notes:

NAME _____ DATE _____

# Lesson 3 Four-Square Vocabulary
## *Regroup a Ten as Ones*

Write the definition for *regroup*. Write what the word means, draw a picture, and write your own sentence using the word.

| **Definition** | **My Own Words** |
|---|---|
| Take apart a number to write it in a new way. | See students' examples. |

**regroup**

| **My Picture** | **My Sentence** |
|---|---|
| See students' drawings. | When I don't have enough ones, I can regroup. |

 **Teacher Directions:** Provide a description, explanation, or example of the new term using images or real objects. Have students use the Glossary to write the definition. Direct students to write a definition in their own words and draw a picture representing their math term. Have students write a sentence using the term. Then encourage students to read their sentence to a peer.

# Lesson 4 Subtract From a Two-Digit Number

## English Learner Instructional Strategy

### Sensory Support: Pictures/Photographs

To introduce the lesson, use pictures of items in groups of 10 to practice counting as a class. Use pictures of the items named in the Lesson/ Problem Solving exercises to clarify the terms: *crabs, beach coins, machine, cafeteria, apples, lunch, store, shirts, and display.*

Pair emerging/expanding students with bilingual (same native language) bridging level students during the Problem Solving exercises. Allow use of native language to clarify meaning.

### English Language Development Leveled Activities

| Emerging Level | Expanding Level | Bridging Level |
|---|---|---|
| **Background Knowledge** | **Building Oral Language** | **Partner Work** |
| Distribute base-ten blocks to students. Write $34 - 8$ on the board. Then use base-ten blocks to model solving the problem with regrouping. Ask, *Can I subtract the ones?* Give students a chance to answer **no**, then say, *No, first I need to regroup.* Have students follow along with their own manipulatives as you finish solving the problem. Write examples and non-examples of two-digit subtraction problems that require regrouping. For each problem, ask, *Do I regroup?* Allow time to answer, confirm answers, and model solving the problem. | On index cards, write examples and non-examples of two-digit subtraction problems that require regrouping. Randomly distribute the cards to students. Have each student explain whether or not regrouping is required. Provide sentence frames, such as: **I need to regroup because _____.** or **I do not need to regroup because _____.** Then have students solve the problems on their cards. For a language extension, discuss the contraction *don't* for *do not* and have students utilize *don't* in the sentence frame. | Have students get into pairs. Have each pair decide which student will be One and which student will be Two. Then say, *Ones will write a two-digit subtraction problem that does not need regrouping. Twos will write a two-digit subtraction problem that needs regrouping.* If necessary, provide an example of each on the board. Then give students time to write their problems. Have the students in each pair exchange problems, solve them, and then provide feedback to each other about whether or not the problem met the requirement. |

**Teacher Notes:**

NAME _____ DATE _____

# Lesson 4 Guided Writing
## *Subtract From a Two-Digit Number*

**How do you subtract from a two-digit number?**

Use the exercises below to help you build on answering the Essential Question. Write the correct word or phrase on the lines provided.

**1.** What key words do you see in the question?

two-digit number _____

| **Word Bank** | | | |
|---|---|---|---|
| ones | regroup | tens | two-digit number |

**2.** A digit is a symbol used to write numbers.
The number 73 is a __two__ - __digit__ __number__.

| tens | ones |
|---|---|
| 6 | 13 |
| 7̸ | 3̸ |
| − 2 | 8 |
| 4 | 5 |

Find 73 − 28.

**3.** Can you subtract the ones? __no__
You must __regroup__ 1 ten as 10 ones.

**4.** Next you should subtract the __ones__.
Then subtract the __tens__.

**5.** How do you subtract from a two-digit number?

Regroup the ones. Subtract the ones. Then subtract the tens. _____

_____

**Teacher Directions:** Read the Building on the Essential Question. Provide descriptions, explanations, or examples of the terms using images or real objects. Read each sentence frame and have students use the lesson and Glossary to write the appropriate terms. Have students read the sentences aloud.

**Grade 2 • Chapter 4** *Subtract Two-Digit Numbers* **35**

# Lesson 5 Subtract Two-Digit Numbers

## English Learner Instructional Strategy

### Sensory Support: Manipulatives/Photos

Before the lesson, have a native language mentor/aide work with students to introduce subtracting two-digit numbers. Have the mentor model and talk students through the 4 steps shown on the See and Show page using Work Mat 6 and base-ten blocks.

To provide additional language support, review the following terms using photos/drawings to clarify the meaning of them: *swimming, pool, diving board, pool pass,* and *lemonade stand* shown in the lesson/Problem Solving exercises.

### English Language Development Leveled Activities

| Emerging Level | Expanding Level | Bridging Level |
|---|---|---|
| **Cooperative Learning** | **Number Game** | **Public Speaking Norms** |
| On the board, write *43 − 18.* Prepare base-ten blocks to use for modeling the problem. Have students form a line. Invite the first student in line to come to the board. Ask, *Do I need to regroup the ones?* Give the student a chance to answer, **yes.** Invite the next student in line to come forward, and to regroup using the manipulatives. Then ask, *How many ones are there now?* **18** Invite the next student to come forward to help subtract the tens. Continue in this manner until the problem is solved. | Pair students. Distribute two (0–5) number cubes and a set of base-ten blocks to each pair. Say, *Each of you roll your number cube. Use the two numbers you each rolled to create a two-digit number.* Have students repeat the process to make a second number. Then tell students to use the two-digit numbers they created to make a subtraction sentence and write it on an index card. Collect the cards. Show the problems one at a time, asking each time, *Do I need to regroup?* Have students answer **yes/no** and then use their manipulatives to solve the problem. | Pair students. Distribute two (0–5) number cubes and a set of base-ten blocks to each pair. Say, *Roll your number cubes and use the two numbers to create a two-digit number. Roll them again to create a second two-digit number. Use your 2 two-digit numbers to create a real-world word problem that requires subtraction to solve.* Give students time to complete the task, and then have pairs share their word problem with the class. Have students work in small groups to solve each word problem. |

**Teacher Notes:**

NAME _____ DATE _____

# Lesson 5 Flow Chart
## *Subtract Two-Digit Numbers*

Read the steps for subtracting two-digit numbers. Write them in correct order in the flow chart.

**Steps for Subtracting**

> Subtract the tens.
>
> Can you subtract the ones?
>
> Subtract the ones.
>
> Regroup I ten as IO ones.

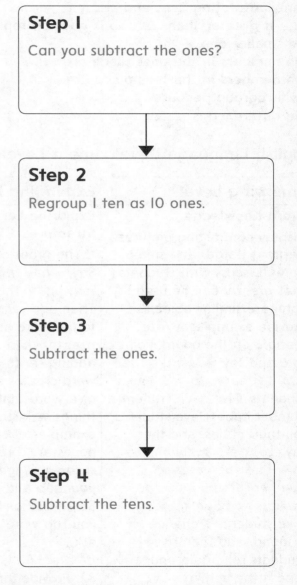

| tens | ones |
|------|------|
| 2̶ 3̶ | 1̶8̶ 8̶ |
| — | 9 |
| 2 | 9 |

**Step I**

Can you subtract the ones?

**Step 2**

Regroup I ten as IO ones.

**Step 3**

Subtract the ones.

**Step 4**

Subtract the tens.

**Teacher Directions:** Read the steps in the box. Review or clarify any unfamiliar terms. Have partners work together to write the steps in order on the flow chart. Then have students switch partners and describe the process to another student. Teach, model, and prompt sequence words such as *first, then, next, last,* and *finally.*

# Lesson 6 Rewrite Two-Digit Subtraction

## English Learner Instructional Strategy

### Vocabulary Support: Communication Guides

Post and model using the following communication guide for students to utilize during the subtraction activities.

I am subtracting _____ and _____.
_____ is greater than _____ so it goes on top of the smaller one.
The numbers in the ones place are . . .
The numbers in the tens place are . . .
I will regroup because . . .
The difference is _____.

## English Language Development Leveled Activities

| Emerging Level | Expanding Level | Bridging Level |
|---|---|---|
| **Word Knowledge** | **Exploring Language Structures** | **Making Connections** |
| Display contrasting pictures of great (large) and small items. Discuss with students that *greater* can be used to compare similar objects. Provide examples. Write 6 and 3 on the board. Point to 6 and say, *6 is more than 3. 6 is **greater** than 3.* Pair students. Give each student a number cube. Have students roll their cubes, and then say, *Look at your numbers. Who has the greater number? If you have the greater number, stand up.* Give students a chance to respond, and then have students roll again and repeat the activity. | Write great on the board. Say, *I had a **great** time playing on the playground.* Then add -er to *great*. Point to the word as you say, *6 is **greater** than 3.* Explain that adding -er to a describing word creates a comparative, or a word that compares two things. Ask students for other examples of comparatives. If necessary, provide examples, such as: *bigger, smaller, younger*, and so on. Write 56 and 32 on the board. Point to word *greater* and ask, *Which number is greater?* **56 is greater than 32.** Provide further examples to extend practice. | Write *greater* and *greatest* on the board. Under *greater*, write the numbers 43 and 28. Under *greatest*, write the numbers 54, 18, and 63. Point to *greater* and say, *Greater ends in -er. It is a comparative, which means it compares two things.* Point to 43 and say, *43 is greater than 28.* Then point to greatest and say, *Greatest ends in -est. It is a superlative, which means it compares more than two things.* Point to 63 and say, *63 is the greatest of these three numbers.* Ask students to provide more examples of comparatives and superlatives. |

## Teacher Notes:

NAME _____ DATE _____

# Lesson 6 Note Taking
## *Rewrite Two-Digit Subtraction*

Read the question. Write words you need help with. Use your lesson to write your Cornell notes. Write or draw math examples to explain your thinking. Share your examples with a classmate.

| **Building on the Essential Question** | **Notes:** |
|---|---|
| How do I rewrite two-digit subtraction? | *Re-* means "again," so *rewrite* means " __write__ __again__ ." |
| | I can __rewrite__ a problem to subtract. |
| | Find 84 − 48. |
| | Write the __greater__ number ⟶ |
| | Write the other number ⟶ __below__ it. |
| **Words I need help with:** | Line up the __ones__ and the __tens__ . |
| See students' words. | Subtract. |
| | If there are not enough __ones__, I should __regroup__ 1 __ten__ as 10 ones. |

Math work:

$$\begin{array}{r} {\scriptstyle 7\ \ 14} \\ \not{8}\ \not{4} \\ 4\ 8 \\ \hline 3\ 6 \end{array}$$

**My Math Examples:**

See students' examples.

**Teacher Directions:** Read the Building on the Essential Question and have students list words/phrases they need assistance with. Provide descriptions, explanations, or examples of the terms using images or real objects. Read each sentence frame and have students write the appropriate terms. Have students read their notes aloud. Direct students to draw a picture representing the question. Then encourage students to describe their picture to a peer.

Grade 2 • **Chapter 4** *Subtract Two-Digit Numbers*  **37**

# Lesson 7 Check Subtraction

## English Learner Instructional Strategy

### Vocabulary Support: Word Webs

Write *fact family* on the board. Review the math term and display an example to support understanding. Accommodate non-Spanish speaking ELs with an appropriate translation tool.

Write *family* at the center of a word web on chart paper. Invite students to add additional words related to *family*. Discuss how the words are related. Then create a word web for *fact family*. Encourage students to add math examples and other vocabulary words such as: *add, addition, subtract, subtraction*, and *related*. Discuss and model that you can check the answer to a subtraction problem by using addition.

### English Language Development Leveled Activities

| Emerging Level | Expanding Level | Bridging Level |
|---|---|---|
| **Phonemic Awareness** | **Phonemic Awareness** | **Share What You Know** |
| Write *ch* on the board. Say the sound /ch/ and have students repeat chorally. Draw a check mark. Say, *Check. This is a **check** mark.* Repeat, asking students to make the mark and say the word **check.** Write a two-digit subtraction problem on the board, such as 45 − 18. Model solving the problem, and then use addition to check the answer. Say, *I will **check** the answer.* Emphasize the /ch/ sound, and have students, who are willing, repeat **check** chorally. | Model solving and checking a two-digit subtraction problem, such as 82 − 35. As you work, say, *I am **checking** my answer.* Emphasize the /ch/ sound and the inflectional ending *-ing*. Write a two-digit subtraction problem on the board, such as 57 − 29. Have students solve the problem. As each student begins adding to check their work, ask them individually, *What are you doing?* Have them say, **I am checking my answer.** Be sure they are correctly saying the /ch/ sound and adding the inflectional ending *-ing*. | Give each student a 6-part spinner numbered 1–6. Have each student spin twice to generate a two-digit number, and then spin two more times to generate a second two-digit number. Say, *Subtract the lesser two-digit number from the greater number.* Give students time to solve the problem. Have students exchange their problems with a partner, and tell them to use addition to check the other student's answer. Then have volunteers explain why addition shows whether the answer is correct or not. |

### Teacher Notes:

NAME _____ DATE _____

# Lesson 7 Guided Writing
## *Check Subtraction*

**How do you check subtraction?**

Use the exercises below to help you build on answering the Essential Question. Write the correct word or phrase on the lines provided.

**1.** What key words do you see in the question?

check, subtraction

**2.** What is the word for the answer to a subtraction problem?

difference

**3.** Add the difference to the number you _subtracted_.

**4.** If your answer is the same as the number you _subtracted_ _from_, your answer is correct.

number you
subtracted from → 42

    −12 ← →Add.

    ——

    30

30 ← difference

+12 ← number you subtracted

——

42 ← number you subtracted from

**5.** How do you check subtraction?

Add the difference to the number I subtracted. The sum should

be the same as the number I subtracted from.

**Teacher Directions:** Read the Building on the Essential Question. Provide descriptions, explanations, or examples of the terms using images or real objects. Read each sentence frame and have students use the lesson and Glossary to write the appropriate terms. Have students read the sentences aloud.

# Lesson 8 Problem Solving
# Strategy: Write a Number Sentence
## English Learner Instructional Strategy

### Collaborative Support: Echo Reading

Pair a native English-speaking student read the problem aloud, and have the EL echo read. Then have students work together using the graphic organizer to solve each problem. Direct students to write a list of "clue" words or phrases from each problem that signaled them to write a subtraction sentence, such as: *fly away, leave, are still, are left, how many more than,* and *eats.* At the end of the lesson, invite students to share their lists of "clue" words.

### English Language Development Leveled Activities

| Emerging Level | Expanding Level | Bridging Level |
|---|---|---|
| **Logical Reasoning** | **Sentence Frames** | **Signal Words** |
| Write and read aloud, *There were 12 birds are in a tree. Then 7 birds flew away. How many birds are left in the tree?* (Provide pictures to clarify word meaning.) Work with students to identify what they know and underline it. Circle the last sentence. Draw a diagram of 12 Xs and then circle 7 of them. Point to Plan, and say, *Now we need a plan. We can use a subtraction number sentence to solve.* Write *number sentence* next to Plan on the board. Encourage students to use the diagram to write a subtraction sentence. | Write then read aloud, *Hector ate 23 grapes. His brother ate 39 grapes. How many more grapes did Hector's brother eat?* Prompt students as they help you solve the problem by asking questions about the four-step problem-solving process. Provide sentence frames for their answers, such as: **We know ____ and ____. We need to find out ____. You should underline ____. You should circle ____. We can use a ____ to solve the problem.** Guide students to use a number sentence as their plan to solve. | Write then read aloud, *There are 52 cards in a deck. Julia dealt 28 cards. How many cards are left in the deck?* Ask students to identify what they know, and then underline the first two sentences of the problem. On the board, write: *who, what, where, when, how,* and *why.* Say, *These are question words. When you see a question word in a problem, it can tell you what you need to find out. Which question word is in the problem on the board?* **how** Have students talk you through the remaining steps and solve the problem. |

**Teacher Notes:**

NAME _____  DATE _____

# Lesson 8 Problem Solving
## STRATEGY: Write a Number Sentence

<u>Underline</u> what you know. (Circle) what you need to find. Write a number sentence to solve.

**1.** <u>There are **25** ants in an ant hill.</u>

<u>**13** ants **leave**.</u>

(How many ants are there **now**?)

___25___ ⊖ ___13___ ⊜ ___12___

There are __12__ ants there **now**.

ant

---

**2.** <u>There are **18** lions in the **yard** at the **zoo**.</u>

<u>**7** lions run into the lion house.</u>

(How many lions are **left** in the **yard**?)

___18___ ⊖ ___7___ ⊜ ___11___

__11__ lions are **left** in the **yard**.

lion

**Teacher Directions:** Provide a description, explanation, or example of the boldface terms and nouns using images or real objects. Read each sentence and have students echo read. Encourage students to write subtraction number sentences and then write their answers in each restated question. Have students read each answer sentence aloud.

**Grade 2 • Chapter 4** Subtract Two-Digit Numbers   **39**

# Lesson 9 Two-Step Word Problems
## English Learner Instructional Strategy

### Collaborative Support: Round the Table

Use a snipping tool to copy then paste each word problem in the lesson onto separate sheets of paper. Place students into multilingual groups of 4 or 5. Distribute one problem from Exercises 1–5 to each group. Have students work jointly on the problem by passing the paper around the table for each member to provide input. Direct each member of the group to write with a different color to ensure all students participate in solving the problem. Provide a step-by-step list for groups to follow, such as: 1) Read the problem aloud as a group and discuss. 2) One student underlines what you know. 3) The next student circles what you need to find. 4) The next student writes a plan. 5) The next student solves the problem. 6) The last student checks for reasonableness. 7) Chose one student to present the solution to the class.

### English Language Development Leveled Activities

| Emerging Level | Expanding Level | Bridging Level |
| --- | --- | --- |
| **Multiple Word Meanings** | **Number Game** | **Cooperative Learning** |
| Invite a boy and a girl to the front of the class. Say, *The difference between ____ and ____ is that ____ is a girl and ____ is a boy.* Have students turn and discuss with a partner the difference between the two students. Support the use of native language for clarification of vocabulary. Discuss as a class. Write 12 and 17 on the board. Model subtracting 12 from 17, and then ask, *What is their difference?* Give students a chance to answer 5, and then say, *Their difference is 5.* Repeat with another pair of numbers. | Pair students. Give one student in each pair a number cube and give the other student a spinner. Write 64 on the board. Tell students to roll their number cubes and spin their spinners. Then say, *Add the number you rolled to the number on the board, and then subtract the number you spun.* Give students time to find the answer. Ask a volunteer to explain the steps using the sentence frames: **First I added ____ to ____. The sum was ____. Then I subtracted ____ from ____. The difference was ____.** Repeat the activity with a new number. | On the board, write the number sentences: 36 − 15 and 21 + 7. Place students into multilingual groups of four. Say, *Write a real-world word problem that uses the two number sentences on the board. Then solve the problem.* Give students time to create their word problem. Circulate among the groups and offer assistance as needed. Then have groups share their word problems and solutions with the entire class. |

### Multicultural Teacher Tip

In some cultures, mental math is strongly emphasized. Latin American students in particular may skip intermediate steps when performing algorithms such as long division. Whereas US students are taught to write the numbers they will be subtracting in the process of long division, Latin American students will make the calculations mentally and write only the results.

NAME _____ DATE _____

# Lesson 9 Guided Writing

## Two-Step Word Problems

### How do you solve a two-step word problem?

Use the exercises below to help you build on answering the Essential Question. Write the correct word from the word bank on the lines provided.

**1.** What key words do you see in the question?

two-step, word problem

> 56 dogs are at the park.
> 14 dogs go home.
> 15 dogs come to the park.
> How many dogs are at the park now?

**Word Bank**
first
know
read
solve

park

**2.** First, I should ___read___ the word problem.

**3.** I should underline what I ___know___.

**4.** Decide what part to solve ___first___.

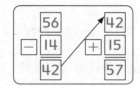

**5.** Then use the rest of the information to ___solve___ the next part.

**6.** How do you solve a two-step word problem?

I would use the information I have, decide which part to solve first, and

then use the rest of the information to solve the next part.

**Teacher Directions:** Read the Building on the Essential Question. Provide descriptions, explanations, or examples of the terms using images or real objects. Read each sentence frame and have students use the lesson and Glossary to write the appropriate terms. Have students read the sentences aloud.

# Chapter 5 Place Value to 1,000

## What's the Math in This Chapter?

### Mathematical Practice 5: Use appropriate tools strategically

Distribute place-value charts, hundred charts, and base-ten blocks to students. Say, *Please use these tools to model the number 97.* Provide time for students to model 97 using each tool and then discuss their strategies and methods as a whole group.

Ask, *What were the similarities and differences between the three tools you used to model 97?* Elicit responses from students such as: **The place value chart used numbers. The base-ten blocks are models. Each model showed 9 tens and 7 ones.** Ask students to share examples of times these tools might be helpful in solving problems. Point out to students that the place-value chart can help write a number in standard and expanded form or when comparing numbers. Base-ten blocks can be helpful to tell how many tens and ones are in different amounts of hundreds. For example, 3 hundreds = 30 tens = 300 ones. The goal of the discussion is for students to recognize that each of these tools might be useful for different situations.

Display a chart with Mathematical Practice 5. Restate Mathematical Practice 5 and have students assist in rewriting it as an "I can" statement, for example: **I can use different math tools to solve problems.** Have students draw or write examples of using math tools to understand place value. Post the examples and new "I can" statement.

## Inquiry of the Essential Question:

### How can I use place value?

Inquiry Activity Target: **Students come to a conclusion that they can use various tools to understand numbers.**

As an introduction to the chapter, present the Essential Question to students. The inquiry graphic organizer will offer opportunities for students to observe, make inferences, and apply prior knowledge of place value representing the Essential Question. As they investigate, encourage students to draw, write, and collaborate with peers to demonstrate their observations and thinking. Then have students present additional questions they may have to a peer to extend discussions.

Regroup students and restate Mathematical Practice 5 and the Essential Question. Pose questions to reflect on what has been learned to guide students in making connections between the Mathematical Practice and the Essential Question.

NAME _____ DATE _____

# Chapter 5 Place Value to 1,000
## *Inquiry of the Essential Question:*

### How can I use place value?

Read the Essential Question. Describe your observations
(I see…), inferences (I think…), and prior knowledge (I know…)
of each math example. Write any questions you have.

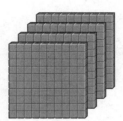

I see …

I think …

I know …

_____ hundreds = _____ tens = _____ ones

---

500 + 70 + 2  =  572

↑ expanded form    ↑ standard form

word form ⟶ five hundred seventy-two

I see …

I think …

I know …

---

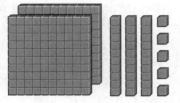

| hundreds | tens | ones |
|----------|------|------|
|          |      |      |

I see …

I think …

I know …

---

Questions I have…

_____

_____

**Teacher Directions:** Read the Essential Question for students. Have students echo read.
Direct students to describe their observations, inferences, and prior knowledge of each
math example. Encourage students to write or draw additional questions they may have.
Then have students share their ideas/questions with a peer.

**Grade 2 · Chapter 5** *Place Value to 1,000* **41**

# Lesson 1 Hundreds

## English Learner Instructional Strategy

### Graphic Support: Number Charts/Manipulatives

Have ELs do the following activity with an aide or peer before the lesson. Review numbers 1 to 50 with verbal counting and visual identification. Then cut the bottom half of a 100 chart into five strips of 10. You will have 51 to 60, 61 to 70, and so on. Use these number lines to review the numbers 51 to 100. Practice saying the numbers with correct pronunciation. Check number recognition by asking students to point to or put manipulatives on certain numbers. Move on to the next number line. Once students show an understanding of all five lines, have them reconstruct the 100 chart.

Introduce and review base-ten blocks for ones, tens, and hundreds before beginning the lesson. Allow students to use them during the Problem Solving exercises as needed.

### English Language Development Leveled Activities

| Emerging Level | Expanding Level | Bridging Level |
|---|---|---|
| **Exploring Language Structure** | **Number Sense** | **Number Recognition** |
| Display 1 ones cube and say, *This is one.* Display 1 tens rod and say, *This is ten.* Display 1 hundreds flat and say, *This is one hundred.* Display 3 tens rods and point to each one in sequence as you say, *one, two, three; three tens.* **three tens** Emphasize the /z/ at the end of *tens.* Display 3 hundreds flats and point to each one in sequence as you say, *one, two, three; three hundreds.* **three hundreds** Emphasize the /z/ at the end of *hundreds.* Randomly display examples of one or more hundreds flats and have students correctly identify the examples. | Write 100 on the board. Display 10 tens rods. Say, *There are ten tens in a hundred.* Display a hundreds flat. Point to a single square. Say, *This is one.* Circle the entire flat with your finger, saying, *There are one hundred ones in a hundred.* Introduce and post sentence frames for students to say how many ones and tens are in a hundred, such as: **There are ____ ones.** **There are ____ tens.** Then ask students, *How many ones in a hundred? How many tens in a hundred?* **100; 10** | Distribute 9 hundreds flats to each student. Display 1 flat and say, *This is a hundred. It is also ten tens. It is also one hundred ones.* Then display 4 flats and say, *This is four hundreds. It is also forty tens. It is also four hundred ones.* Say, *Show me three hundreds.* Have each student display three hundreds flats. Write 300 on the board. Have a volunteer use the sentence frames: **This is ____ hundreds.** **It is ____ tens.** **It is also ____ ones.** Repeat for all other multiples of a hundred through 900. |

### Teacher Notes:

NAME _____     DATE _____

# Lesson I Concept Web
## *Hundreds*

Use the concept web to show different ways to write hundreds.

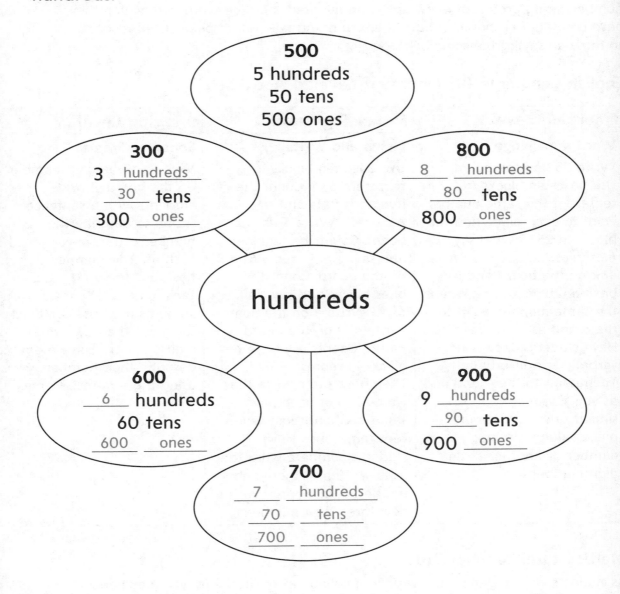

**500**
5 hundreds
50 tens
500 ones

**300**
3 ___hundreds___
___30___ tens
300 ___ones___

**800**
___8___ hundreds
___80___ tens
800 ___ones___

**hundreds**

**600**
___6___ hundreds
60 tens
___600___ ones

**900**
9 ___hundreds___
___90___ tens
900 ___ones___

**700**
___7___ hundreds
___70___ tens
___700___ ones

**Teacher Directions:** Review the terms *hundreds, tens,* and *ones.* Model and prompt these sentences such as **400 has 4 hundreds. That equals 40 tens. It also equals 400 ones.** Have partners complete the chart. Then have students switch partners and compare answers with another student.

# Lesson 2 Hundreds, Tens, and Ones
## English Learner Instructional Strategy

### Graphic/Sensory Support: Work Mats/Manipulatives

Have ELs do the following activity with an aide or older peer before the lesson. Provide students a copy of Work Mat 7: Hundreds, Tens, and Ones Chart and base-ten blocks. Have them available for students to use during the Problem Solving exercises as well. Allow students to experiment creating their own numbers using groups of ones, tens, and hundreds on their work mat. When they have created their number, have them turn and talk with a peer or report back to the aide saying their number aloud.

### English Language Development Leveled Activities

| Emerging Level | Expanding Level | Bridging Level |
|---|---|---|
| **Word Knowledge** | **Listen and Write** | **Sentence Frames** |
| Write 253 on the board. Use base-ten blocks to represent the same number. Point to each place and the blocks in 253 as you say, *hundreds, tens, ones.* Write 462 on the board and use base-ten blocks to represent the same number. Point to the 6 and ask, *hundreds?* Have students answer **no** verbally or nonverbally. Then point to the 4 and nod as you say, *hundreds.* Ask similar questions about place value using this number or another three-digit number. | Use base-ten blocks to represent 584. Count the hundreds flats and say, *five hundreds.* Write 5 on the board. Count the tens rods and say, *eight tens.* Write 8 on the board. Count the ones blocks and say, *four ones.* Write 4 on the board. Write 679 on the board. Have students use base-ten blocks to represent 679. Then have them write the word form of each place value. **six hundreds, seven tens, nine ones** Have students compare what they have written to the number on the board. Repeat with a new three-digit number. | Write a three-digit number on the board. Provide students with a sentence frame to describe the number you have written, such as: **The number has _____ hundreds, _____ tens, and _____ ones.** Have students use base-ten blocks to create their own three-digit numbers. Have them describe their numbers using the sentence frame. |

### Multicultural Teacher Tip

As students work together in a variety of collaborative situations, you may notice some EL students seem reluctant to participate. This behavior could be the result of shyness or insecurity due to language issues, but the student may also come from a classroom environment in his or her native culture that did not emphasize group work. The student may be unsure of how to participate or what role to play, worrying that collaboration may be akin to cheating. Be patient and encouraging, and allow the student time to get comfortable with working in a group dynamic.

NAME _____ DATE _____

# Lesson 2 Vocabulary Sentence Frames

*Hundreds, Tens, and Ones*

The math words in the box are for the sentences below. Write the words that fit in each sentence on the blank lines.

| hundreds | tens | ones |
|----------|------|------|

**I.** The numbers in the range of 0 to 9 are called ___ones___.

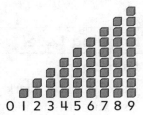

**2.** 50 tens equal 5 ___hundreds___.

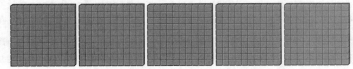

**3.** In the number 45, there are 4 ___tens___ and 5 ones.

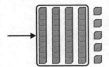

 **Teacher Directions:** Provide a description, explanation, or example of the each term using images or real objects. Read each sentence frame and have students echo read. Direct students to write the correct terms in each blank. Then encourage students to read each sentence to a peer.

Grade 2 • **Chapter 5** *Place Value to 1,000* **43**

# Lesson 3 Place Value to 1,000

## English Learner Instructional Strategy

### Vocabulary Support: Frontload Academic Vocabulary

Before the lesson, write *place value, expanded form* and *digit (dígito)* on a chart. Introduce the words, write a math example, and model using concrete objects (when appropriate) to support understanding.

Have students use index cards to write the terms and draw a visual example. Using the Glossary, have students write definitions and then sentences using each term in context. Provide sentence examples on the chart if necessary. Accommodate non-Spanish speaking ELs with an appropriate translation tool. Have students glue a pocket into their math journals to store cognate cards for future review.

### English Language Development Leveled Activities

| Emerging Level | Expanding Level | Bridging Level |
|---|---|---|
| **Choral Responses** | **Number Recognition** | **Pairs Share** |
| Using base-ten blocks, show 1 ones cube, 1 tens rod, and 1 hundreds flat. Model aloud saying the place values, both sequentially and in reverse. Have students chorally repeat. Emphasize the sound /z/ at the end of the place values: ones, tens, and hundreds. Next, randomly display each example. Have students name each example as it is displayed. Model 315. Say the expanded form, *three hundreds, one ten, five ones.* Students repeat, **three hundreds, one ten, five ones.** Repeat with other three-digit numbers up to 1,000. | Write a series of one-digit numbers on the board, such as: 8, 1, 5, 2. Allow extra space between the numbers. Read each number aloud. Next, add digits so that each number is a one-, two-, or three-digit number, such as: 18, 217, 59, 412. Ask students to read the new numbers. | Have students work in pairs. Ask students to identify a new or unique way to model the three-digit number 458. Provide them a choice of multiple concrete objects to represent different place values. As a group, discuss why each group's model represents 458. Ask what they might do to model a greater number. |

**Teacher Notes:**

NAME _____ DATE _____

# Lesson 3 Word Identification
## *Place Value to 1,000*

Match each word to the example.

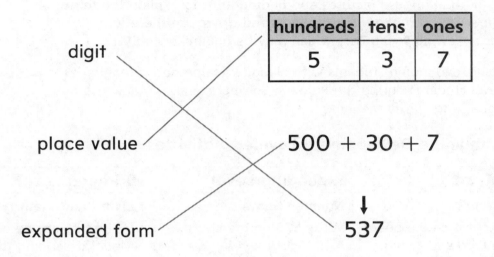

| hundreds | tens | ones |
|----------|------|------|
| 5 | 3 | 7 |

digit

place value

500 + 30 + 7

↓

537

expanded form

---

Write the correct word from above for each sentence on the blank lines.

1.  The ___expanded___ ___form___ shows the value of each digit.

2.  A ___digit___ is a symbol for a number.

3.  In 537, the ___place___ ___value___ of the 5 is 5 hundreds.

**Teacher Directions:** Review the terms using images or real objects. Have students say each term and then draw a line to match the term to its meaning. Direct students to say each sentence and then write the corresponding meanings in the sentences. Have partners compare answers and read their sentences to each other.

# Lesson 4 Problem Solving Strategy: Use Logical Reasoning

## English Learner Instructional Strategy

### Sensory Support: Realia

Prior to the lesson, review the singular and plural words: *penny/pennies* and *cherry/cherries* by sharing the "real" objects with students to support the lesson. In addition, use pictures and photographs to explain the terms: *house number, dogs, shelter, cousins, visit,* and *drive.* Use these for teaching or reviewing vocabulary which may be unfamiliar to ELs.

Pair emerging/expanding students with bilingual (same native language) bridging level students during the Problem Solving exercises. Allow use of native language to clarify meaning.

## English Language Development Leveled Activities

| Emerging Level | Expanding Level | Bridging Level |
|---|---|---|
| **Number Sense** Draw a T-chart on the board and write *Odd* and *Even* as heads. Below Odd, list odd numbers 1 to 9. Below Even, list even numbers 2 to 10. Write 49 on the board. Point to 9 in 49 and then 9 in the list. Say, *This number is **odd.*** Write 32 on the board. Point to 2 in 32 and then 2 in the list. Say, *This number is **even.*** Write another two-digit number on the board. Point to the number in the ones place and ask, *Is this number odd or even?* Have students answer accordingly. Repeat with additional two- and three-digit numbers. | **Number Game** Write 578 on the board. Ask, *Which number is in the ones place?* **8** Then write 254. Ask, *Which number is in the hundreds place?* **2** Tell students they can solve problems using place value. Say, *I am thinking of a three- digit number.* Have students draw three short blank spaces. Say, *My number has a 7 in the tens place, a 4 in the hundreds place, and a 2 in the ones place.* Check that all students have written 472. Repeat allowing volunteers to describe a three-digit number. | **Listen and Identify** Write 534 on the board. Model describing the number using math vocabulary, *The first digit is odd. The middle digit is 3. The digit in the ones place is 4. The smallest digit is in the middle.* Write other three-digit numbers on the board. Have students take turns saying sentences that describe the numbers. Draw three blank spaces on the board. Say, *The digit in the hundreds place is an odd number. It is less than three. The middle digit is the sum of 3 and 5. The last digit is the greatest.* Guide students as they identify the number as 189. |

**Teacher Notes:**

NAME _____ DATE _____

# Lesson 4 Problem Solving
## *STRATEGY: Use Logical Reasoning*

<u>Underline</u> what you know. (Circle) what you need to find. Use logical reasoning to solve the problems.

**I.** <u>A number has three **digits**.</u>

<u>The digit in the **hundreds** place is the **difference** of 4 and 2.</u>

| hundreds | tens | ones |
|:---:|:---:|:---:|
| 2 | 5 | 7 |

<u>The digit in the **tens** place is the **sum** of 2 and 3.</u>

<u>The digit in the **ones** place is 2 **more** than the digit in the tens place.</u>

(What is the number?)

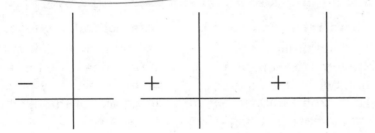

The number is __257__.

---

**2.** <u>Sylvia writes this number: **three ones, two tens,** and **seven hundreds**.</u>

| hundreds | tens | ones |
|:---:|:---:|:---:|
| 7 | 2 | 3 |

(What number is it?)

The number is __723__.

**Teacher Directions:** Provide a description, explanation, or example of the boldface terms and nouns using images or real objects. Read each sentence and have students echo read. Encourage students to use logical reasoning and what they know about place-value charts to figure out the answer. Then have them write their answers in each restated question. Have students read each answer sentence aloud.

Grade 2 • **Chapter 5** *Place Value to 1,000* **45**

# Lesson 5 Read and Write Numbers to 1,000

## English Learner Instructional Strategy

### Vocabulary Support: Communication Guide

Post and model using the following communication guide for students to utilize during the place value activities.

**The missing number is _____.**
**The digit in the ones place is _____.**
**_____ is the digit in the tens place.**
**The digit in the hundreds place is _____.**
**The number is _____.**

Invite students to record the sentence frames in a journal and use them to write complete sentences using the numeral and word forms of three-digit numbers.

### English Language Development Leveled Activities

| Emerging Level | Expanding Level | Bridging Level |
|---|---|---|
| **Non-Transferable Sounds** | **Share What You Know** | **Internalize Language** |
| On the board, write the numbers and words for: 3, 13, 30, and 1,000. Pointing to each one, say the number word emphasizing the correct /th/ sound and have students repeat. Point to the words, and have students identify them aloud. Check for correct pronunciation of the /th/ sound. Students may interchange /t/ or /tw/. Model tongue position for /th/ as needed. | In small groups, ask students to think of a number in the hundreds. Each group will demonstrate the number by using manipulatives, acting it out, or creative nonlinguistic representations. Have each group demonstrate their number. Other students will guess the number by writing their guesses and displaying them. The rest of the group show thumbs-up if they think it is correct, or thumbs-down if they think it is incorrect. | Write 4 three-digit numbers on the board. Tell a real-world, short story problem using one of the numbers. For example, *There are _____ pieces of popcorn in the bowl. Say the digit in the tens place.* Have students take turns telling stories and answering the posed place value question using each number. |

### Teacher Notes:

NAME _____ DATE _____

# Lesson 5 Word Web

## *Read and Write Numbers to 1,000*

Write each number in words. Then read the number.

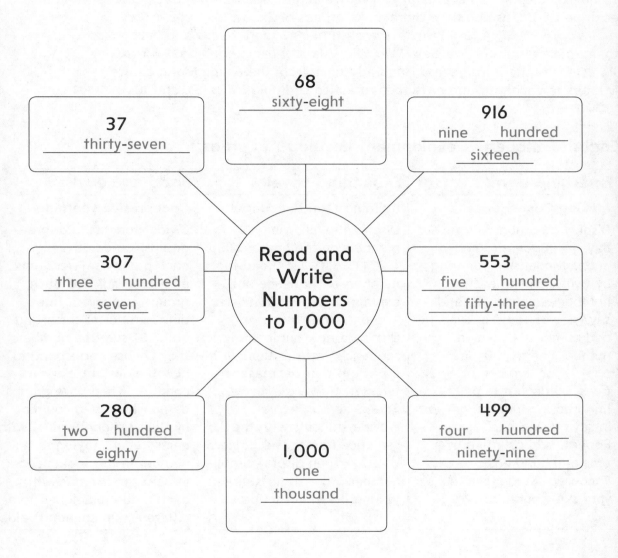

68
_____ sixty-eight

37
_____ thirty-seven

916
_____ nine _____ hundred
_____ sixteen

307
_____ three _____ hundred
_____ seven

553
_____ five _____ hundred
_____ fifty-three

**Read and Write Numbers to 1,000**

280
_____ two _____ hundred
_____ eighty

1,000
_____ one
_____ thousand

499
_____ four _____ hundred
_____ ninety-nine

**Teacher Directions:** Review the spellings and pronunciations of the ones, teens, tens, and hundreds using the chart on student page 322. Give extra pronunciation practice for words that include the /th/ sound, which is often difficult for English learners. Also review words with unusual or unexpected spellings, such as *eight* or *forty*. Remind students as they write a number that has tens and ones, they should use a hyphen to separate the two parts (as in forty-eight).

# Lesson 6 Count by 5s, 10s, and 100s

## English Learner Instructional Strategy

### Sensory Support: Manipulatives

Review the term *skip count*. Refer to the Math Word Wall and remind students that they have practiced skip counting by twos.

To quickly review skip counting, pair native English speakers with ELs. Give each pair 8 beans and have them make groups of 2. Say, *Skip count these beans with me:* **2, 4, 6, 8.** Next, give each pair 12 more beans and have them make groups of five. Say, *Skip count the beans with me again:* **5, 10, 15, 20** Distribute a Numbered Hundred Chart, from the online Manipulative Masters to each pair and have them practice counting by 2s, 5s, and 10s to 100.

### English Language Development Leveled Activities

| Emerging Level | Expanding Level | Bridging Level |
|---|---|---|
| **Making Connections** | **Building Oral Language** | **Cooperative Learning** |
| Display a number line to 10. Say, *I am going to skip count.* Model skip counting by twos. Point to each number as you say it, and say, *skip* as you skip over the odd numbers. Display a number line to 100. Say, *Now let us skip count by fives.* Skip count to one hundred as students join you in saying **skip** each time. Repeat, skip counting by tens and hundreds. Encourage students to echo your skip counting. | Display the following numbers on the board: 125, 130, 135, 140, 145. Read the numbers in sequence and explain, *There is a **pattern**. The pattern is **five more**.* Display a new number sequence, but this time with a pattern that decreases by ten. Say, *There is a **pattern**. The pattern is **ten less**.* Provide students with the sentence frame: **The pattern is ____.** Then display simple number patterns for students to identify and describe. | Divide students into three groups. Randomly assign each group as: Fives, Tens, and Hundreds. Tell each group to choose a three-digit number as a starting point. Be sure the numbers they choose correspond to how they will be skip counting. Then have groups demonstrate skip counting by their assigned value, with each student saying a number in the sequence. Repeat the exercise with each group reassigned to a different skip counting value. |

**Teacher Notes:**

NAME _____ DATE _____

# Lesson 6 Guided Writing

*Count by 5s, 10s, and 100s*

## How do you count by 5s, 10, and 100s?

Use the exercises below to help you build on answering the Essential Question. Write the correct word from the word bank on the lines provided.

**1.** What key words do you see in the question?

count by _____

---

### Word Bank

counting pattern    fives    hundreds    patterns    tens

---

**2.** Number __patterns__ can help you count more or count less.

**3.** When you <u>count</u> <u>on</u> by __fives__, each number is 5 **more**.

**Example: 45, 50, 55, 60, 65**

**4.** When you <u>count</u> <u>back</u> by __tens__, each number is 10 **less**.

**Example: 45, 35, 25, 15, 5**

**5.** When you <u>count</u> <u>on</u> by __hundreds__, each number is 100 **more**.

**Example: 45, 145, 245, 345, 445**

**6.** The __counting__ __pattern__ for 433, 333, 233, 133 is 100 **less**.

**7.** How do you count by 5s, 10s, and 100s?

Sample answer: I can count using a number pattern, such as 10 more or

5 less.

**Teacher Directions:** Read the Building on the Essential Question. Provide descriptions, explanations, or examples of the terms using images or real objects. Read each sentence frame and have students use the lesson and Glossary to write the appropriate terms. Have students read the sentences aloud.

**Grade 2 • Chapter 5** *Place Value to 1,000* **47**

# Lesson 7 Compare Numbers to 1,000
## *English Learner Instructional Strategy*

### Language Structure Support: Modeled Talk

Have ELs do the following activity with an aide or older peer before the lesson. Give each pair three 0–5 number cubes. Have them roll the cubes, and use the three numbers to make two different three-digit numbers and write them in their math journals. If, for example, the cubes land as 2, 5, and 5, students could write 252 and 552. Have them report back comparisons between the numbers. Teach and model the following sentence frames for students to use during reporting:

Emerging students: Ask, *Is 252 greater than or less than 522?* **less than**

Expanding students: \_\_\_\_ **is greater/less than** \_\_\_\_.

Bridging students: \_\_\_\_ **is greater/less than** \_\_\_\_ **because** \_\_\_\_.

### English Language Development Leveled Activities

| Emerging Level | Expanding Level | Bridging Level |
|---|---|---|
| **Number Sense** Roll two number cubes. Write and introduce sentence frames to compare the numbers on each cube, such as: \_\_\_\_ **and** \_\_\_\_ **are equal.** \_\_\_\_ **is greater than** \_\_\_\_. \_\_\_\_ **is less than** \_\_\_\_. Write *compare*. Explain to students that you just compared two numbers. Elicit students to say the term *compare numbers*. **compare numbers** Give pairs number cubes and have them repeat the activity using the sentence frames. | **Exploring Language Structure** Write two numbers on separate sticky notes. Place the notes next to each other. Say, *I can compare the numbers.* Then compare each number, using the symbols <, >, or =. Say, *I am comparing two numbers.* Write each form of compare, underlining the inflectional ending -ing. Have students repeat the activity, using their own numbers. | **Building Oral Language** Have pairs use a number line or another visual model of their choice to compare 2 three-digit numbers. Introduce sentence frames to help students compare the numbers, such as: \_\_\_\_ **is greater than/less than/ equal to** \_\_\_\_ **because** \_\_\_\_. Students will use their knowledge of ones, tens, and hundreds to explain their answers. |

### Teacher Notes:

NAME _____ DATE _____

# Lesson 7 Note Taking

## *Compare Numbers to 1,000*

Read the question. Write words you need help with.
Use your lesson to write your Cornell notes. Write or
draw math examples to explain your thinking.

| **Building on the Essential Question** | **Notes:** |
|---|---|
| How can I use a place-value chart to compare numbers to 1,000? | Write the numbers being ordered on a place-value <u>chart</u>.<br><br>table below<br><br>Compare the digits in the greatest <u>place</u> <u>value</u>.<br><br>If the digits are the same, <u>compare</u> the digits in the next place value.<br><br>Continue to compare until the digits are <u>different</u>.<br><br>Then compare using <u>greater</u> <u>than</u> (>), <u>less</u> <u>than</u> (<), or equal to (=).<br><br>732 ⓒ 737 |

Place-value chart:

| hundreds | tens | ones |
|---|---|---|
| 7 | 3 | 2 |
| 7 | 3 | 7 |

**Words I need help with:**
See students' words.

**My Math Examples:**
See students' examples.

 **Teacher Directions:** Read the Building on the Essential Question and have students list words/phrases they need assistance with. Provide descriptions, explanations, or examples of the terms using images or real objects. Read each sentence frame and have students write the appropriate terms. Have students read their notes aloud. Direct students to draw an example representing the question and describe it to a peer.

# Chapter 6 Add Three-Digit Numbers
## What's the Math in This Chapter?

### Mathematical Practice 8: Look for and express regularity in repeated reasoning

Distribute base-ten blocks to pairs of students and write the addition problem 316 + 472 vertically on the board. Have each pair member model one of the numbers using their base-ten blocks. Once each pair has a model representing 316 and a model representing 472, ask students to combine their like pieces. Say, *How many ones, tens, and hundreds are in each pile?* **7 hundreds, 8 tens, and 8 ones.** Record on the board.

Discuss with students how this is similar to adding two-digit numbers together. They added like place-value together. Model solving the problem on the board in a place-value chart. Highlight the connection between the answer that they found with their base-ten blocks to the traditional algorithm. Remind students they can use the same algorithm repeatedly with any addition problem, beginning with the ones place and regrouping when needed.

Display a chart with Mathematical Practice 8. Restate Mathematical Practice 8 and have students assist in rewriting it as an "I can" statement, for example: **I can notice when I need to repeat calculations to solve a problem.** Post the new "I can" statement.

## Inquiry of the Essential Question:

### How can I add three-digit numbers?

Inquiry Activity Target: **Students come to a conclusion that they can use repeated calculations to add larger numbers.**

As an introduction to the chapter, present the Essential Question to students. The inquiry graphic organizer will offer opportunities for students to observe, make inferences, and apply prior knowledge of the addition process representing the Essential Question. As they investigate, encourage students to draw, write, and collaborate with peers to demonstrate their observations and thinking. Then have students present additional questions they may have to a peer to extend discussions.

Regroup students and restate Mathematical Practice 8 and the Essential Question. Pose questions to reflect on what has been learned to guide students in making connections between the Mathematical Practice and the Essential Question.

NAME _____ DATE _____

# Chapter 6 Add Three-Digit Numbers
## *Inquiry of the Essential Question:*

**How can I add three-digit numbers?**

Read the Essential Question. Describe your observations
(I see...), inferences (I think...), and prior knowledge (I know...)
of each math example. Write any questions you have.

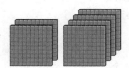

> **THINK**
> You know that $2 + 4 = 6$.
> So, $200 + 400 = 600$.

I see ...

I think ...

I know ...

$$2 \text{ hundreds}$$
$$+ \ 4 \text{ hundreds}$$
$$\overline{6 \text{ hundreds}}$$

---

**Mentally add 100.**

$$530$$
$$+ 100$$

> **THINK** $5 + 1 = 6$, so $\begin{array}{r} 530 \\ + 100 \\ \hline 630 \end{array}$

I see ...

I think ...

I know ...

**Mentally add 10.**

$$565$$
$$+ \ 10$$

> **THINK** $6 + 1 = 7$, so $\begin{array}{r} 565 \\ + \ 10 \\ \hline 575 \end{array}$

---

Questions I have...

_____

_____

**Teacher Directions:** Read the Essential Question for students. Have students echo read.
Direct students to describe their observations, inferences, and prior knowledge of each
math example. Encourage students to write or draw additional questions they may have.
Then have students share their ideas/questions with a peer.

# Lesson 1 Make a Hundred to Add

## English Learner Instructional Strategy

### Collaborative Support: Native Language Peers/Mentors

Review the term *take apart* with students. Explain they will take apart numbers to make a hundred to make it easier to add. Use the online base-ten blocks, in Virtual Manipulatives to assist students visually as they model along with you during the Explore and Explain activity.

Pair bridging level students with same native language emerging/expanding students. Have the pairs work together throughout the lesson. Invite them to explain their thinking with the sentence frames: **I took apart _____ to get _____ + _____. I added _____ and _____ to make a hundred. Then I added _____ + _____ to get the sum, _____.**

### English Language Development Leveled Activities

| Emerging Level | Expanding Level | Bridging Level |
|---|---|---|
| **Show What You Know** Use manipulatives to model $98 + 8 =$ _____. Explain that the problem will be easier if you first add to make a hundred. Say, *I will add a number to 98 to make it 100. What number do I add?* Write $98 + 2 = 100$ on the board. Repeat, *I add 2 to make a hundred. Now I take away the 2 to get the correct answer. I take apart the eight. $8 - 2 = 6$. Now the problem is $100 + 6 =$* _____. *The answer is 106.* Repeat with $87 + 14 =$ _____. Allow students to work through examples with manipulatives to indicate how they add and take away to get the answer. | **Exploring Language Structures** Write $198 + 37 =$ _____. Write each step as you say, *I can make a hundred to add. First I take apart 37. Next I add 2 to 198 to make 200. Then I add 200 to 35 to find the sum 235.* Tell students to listen carefully, because you are going to explain the steps again. *First I took apart 37. Next I added 2 to 198 to make 200. Then I added 200 to 35 to find the sum 235.* Stress the -ed ending in the word *added*. Write $96 + 8 =$ _____. Have students solve by making a hundred. Then have a volunteer use these sentence frames: **First I took apart _____. Next I added _____ to _____ to make _____. Then I added _____ to _____ to find the sum _____.** | **Public Speaking Norms** Distribute base-ten blocks to pairs of students. Write $294 + 9 =$ _____. Have pairs of students work together to solve the problem by making a hundred to add. Have one student in each pair explain how they solved the problem. Make sure students use *add* and *hundred* in their explanations by giving them the sentence frame: **I made a hundred by adding _____ and _____. Then I added _____ and _____ to get the sum _____.** |

## Teacher Notes:

## Lesson I Word Web

### *Make a Hundred to Add*

Read the math words. Draw a picture or write an example in each rectangle that shows the meanings of *addend* and *sum*. See students' examples.

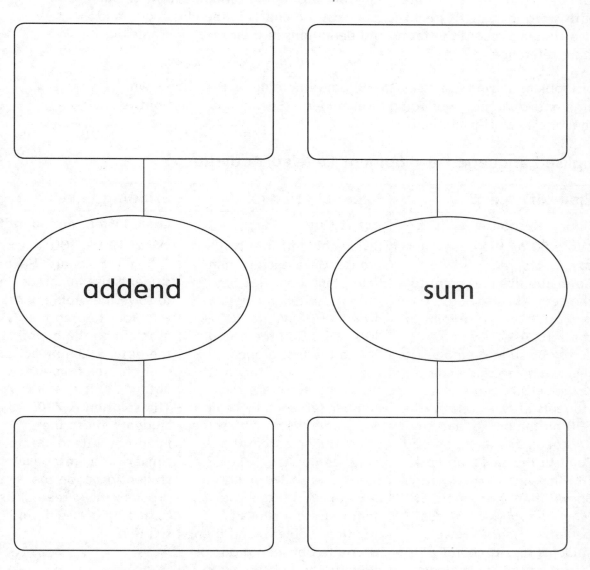

 **Teacher Directions:** Provide a description, explanation, or example of the new terms using images or real objects. Have students say the term aloud. Direct students to draw pictures or write examples that represent the math terms. Then encourage students to describe their work to a peer.

# Lesson 2 Add Hundreds
## English Learner Instructional Strategy

### Vocabulary Support: Activate Prior Knowledge

Use the Math Word Wall to review the terms: *hundred, subtraction, difference,* and *zero.* Organize students into groups of 3–5 students and assign a term to each group. Have students look up the term in the Glossary. Provide students with chart paper to write a definition, and draw a visual math example with the term labeled. Then, invite a volunteer from each group to share their definitions using the sentence frames: **Our math word is ____. It means ____.** Post the charts in the classroom, and have students record the terms and definitions in their math journals for future reference.

For Problem Solving Exercises 26–28, pair emerging level students with same native language bridging level students. Encourage use of native language to clarify meaning.

### English Language Development Leveled Activities

| Emerging Level | Expanding Level | Bridging Level |
|---|---|---|
| **Choral Response** Write 50 on the board, and say, *This number is fifty.* Cover the five and point to the zero while saying, *This number has a zero.* Point to the zero. Ask, *How many zeros?* Be sure to emphasize the beginning and ending /z/ sounds of zeros. Have students answer, **one.** Write 500 on the board, and say, *This number is five hundred.* Point to the zeros, and ask, *How many zeros?* **two zeros** Show 5 flats and 2 flats. Say, *5 hundred plus 2 hundred equals 7 hundred.* Have students repeat chorally. Distribute flats to students and repeat with similar problems. | **Act It Out** Divide students into two groups. Have each student hold a hundreds flat. Say, *Each person in your group is equal to 100.* Write 200 + 300 = ____ on the board. Say to the first group, *Show me 200 + 300.* Students should demonstrate the number sentence by having two students join with three other students from the same group. Ask, *What is the solution?* **five hundred** Ask the second group to demonstrate a similar addition problem, and then alternate between groups directing and observing to demonstrate number sentences that add hundreds. | **Look, Listen, and Identify** Write 400 + 300 = ____ on the board. Have students use hundreds flats to solve the problem. Ask a volunteer to read the problem on the board, state the basic addition fact and then the solution. **400 + 300 = ____; 4 + 3 = 7; The solution is 700.** Ask students to use the hundreds flats to create similar problems. Have each student describe the problem they created, its basic addition fact, and its solution. |

### Teacher Notes:

NAME _____ DATE _____

# Lesson 2 Vocabulary Word Study
## *Add Hundreds*

Circle the correct word to complete the sentence.

The greatest <u>place</u> <u>value</u> of the numbers in the range of 100 to 999 is hundreds.

expanded form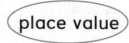

---

Show what you know about the word:
# hundreds

There are __8__ letters.

There are __2__ vowels.

There are __6__ consonants.

__2__ vowels + __6__ consonants = __8__ letters in all.

---

Draw a picture to show what the word means.

See students' examples.

**Teacher Directions:** Provide a description, explanation, or example of the new term using base-ten blocks or real objects. Read the sentence and have students circle the correct word. Direct students to count the letters, vowels and consonants in the math term, then complete the addition number sentence. Guide students to draw a picture representing their math term. Then encourage students to describe their picture to a peer.

Grade 2 • **Chapter 6** *Add Three-Digit Numbers* **51**

# Lesson 3 Mentally Add 10 or 100
## English Learner Instructional Strategy

### Sensory Support: Pictures and Photographs

To introduce the lesson, use pictures of items in groups of 10 and 100 to practice counting as a class. Use pictures of the items named in the Lesson/Problem Solving activities to clarify the terms: *bus*, *birds*, *park*, *mail carrier*, *delivered*, *package*, *sheep*, and *barn* to provide additional language support.

Pair emerging/expanding students with bilingual (same native language) bridging level students during the Problem Solving exercises. Allow use of native language to clarify meaning.

### English Language Development Leveled Activities

| Emerging Level | Expanding Level | Bridging Level |
|---|---|---|
| **Number Sense** | **Sentence Frames** | **Developing Oral Language** |
| Have students write *Tens* on one index card and *Hundreds* on a second index card. Write 145 + 200 = ____. Ask, *Should I mentally add the hundreds place or the tens place?* Students should display the *Hundreds* card. Then model adding mentally to get the answer. Say, *I know one plus two is three, so I can add the hundreds in my mind.* Write 356 + 10 = ____. Ask, *Should I add the hundreds place or the tens place?* Students should display the *Tens* card. Then have them add mentally to find the answer to the problem. Repeat with similar problems. | Write 1 + 3 = ____ and 100 + 300 = ____. Introduce a sentence frame to help students find the answer: ____ **plus** ____ **equals** ____, **so** ____ **hundred plus** ____ **hundred equals** ____ **hundred.** Model using the sentence frame by saying, *One plus three equals four, so one hundred plus three hundred equals four hundred.* Write other hundreds problems for students to answer using the sentence frame. | Write 235 + 300 = ____, and underline the numbers in the hundreds places. Say, *If two plus three equals five, then two hundred plus three hundred equals five hundred.* Model solving the problem first mentally then on the board using this information. Write 418 + 200 = ____. Underline the numbers in the hundreds places. Have students use an, **If** ____, **then** ____, conditional phrase to demonstrate adding mentally to find the answer. Repeat with other addition problems. |

**Teacher Notes:**

NAME _____ DATE _____

# Lesson 3 Concept Web
## *Mentally Add 10 or 100*

Use the concept web to show different ways to
mentally add 10 or 100.

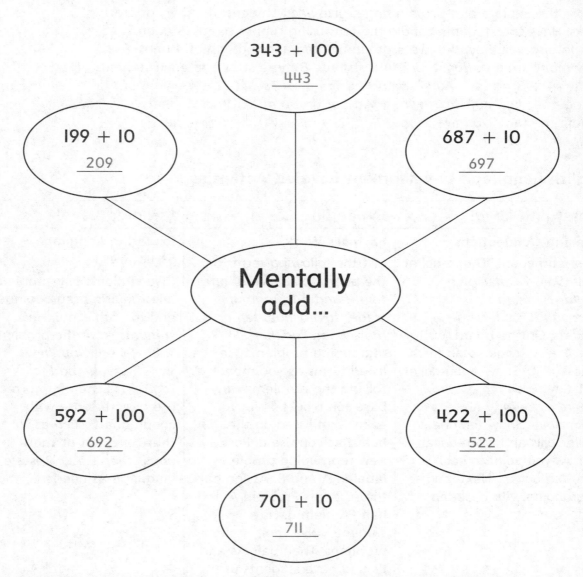

343 + 100
443

199 + 10
209

687 + 10
697

Mentally add...

592 + 100
692

422 + 100
522

701 + 10
711

**Teacher Directions:** Review the terms *hundreds* and *tens*. Then tell students that
*mentally add* means to add in their heads. Before students begin the activity, have
partners read the expressions to each other. Have partners complete the chart. Then
have students switch partners and compare answers with another student using the
sentence frame: _____ **plus** _____ **equals** _____.

# Lesson 4 Regroup Ones to Add
## English Learner Instructional Strategy

### Sensory Support: Act It Out

Write *regroup* and its Spanish cognate, *reagrupar*, on the cognate chart. Introduce the word, and model the following activity to support under standing. Assist non-Spanish speaking ELs with an appropriate translation tool.

Before the Explore and Explain page, give each student 14 straws or craft sticks. Have the students bundle the tens using rubber bands. Students should discover they all have a group of 1 ten and a group of 4 ones. Ask, *How many straws altogether?* **14** Distribute 8 more straws to each student. Write 14 + 8 =. Say, *When we add 8 to our 4 we get 12. We can regroup 10 together and put them with our other group of 10.* Have students bundle the ten. *We have 2 ones left. How many straws do we have altogether?* **22**

### English Language Development Leveled Activities

| Emerging Level | Expanding Level | Bridging Level |
|---|---|---|
| **Making Connections** | **Partners Work** | **Exploring Language Structures** |
| Post and count 10 one-dollar bills. Say, *1 dollar plus 1 dollar . . . (and so on through 10 dollars) makes 10 dollars.* On the board, write 1 + 1 + . . . (and so on for 10 ones) = 10. Show a ten-dollar bill. Say, *Each ten-dollar bill is like a group of ten one-dollar bills.* Show ten one dollar bills and a ten-dollar bill. As you point to each, say, *ten dollars.* Next model regrouping with base-ten blocks. | List the following terms on the board: *place, add, ones, tens, hundreds, regroup, trade, sum, value, ten or more ones.* Post a regrouping problem. Use the listed terms as you model solving the problem with base-ten blocks. Point to each term listed on the board as you use it. Post 2 new regrouping problems labeled A and B. Assign half the students Problem A and half Problem B. Have students solve their problems. Then pair As and Bs and have students use the listed terms to explain their solutions. | Give students two problems, one requiring regrouping and the other not. Have students solve the problems. Ask, *How do you know whether you need to regroup?* Listen for correct use of verb tense with conditionals, such as, ***If*** there *are* **ten or more ones, *then* _____.** Restate language as needed. |

### Teacher Notes:

NAME _____ DATE _____

# Lesson 4 Flow Chart
## *Regroup Ones to Add*

Read the steps for regrouping ones to add with three-digit numbers. Fill in the missing words on the blank lines. Use your lesson to help.

---

**Step 1**

Add the ___ones___. If there are 10 or ___more___ ones, ___regroup___ 10 ones as 1 ten. Write the 1 in the tens ___column___.

---

**Step 2**

Add the ___tens___.

---

**Step 3**

Add the ___hundreds___.

---

 **Teacher Directions:** Read the steps in the flow chart. Review or clarify any unfamiliar terms. Have partners use the lesson and work together to write the missing terms in the sentence frames. Have students find 412 + 259 and describe the process to their partner using the flow chart.

**Grade 2 • Chapter 6** *Add Three-Digit Numbers* **53**

# Lesson 5 Regroup Tens to Add
## English Learner Instructional Strategy

### Vocabulary Support: Communication Guide

Post and model using the following communication guide for students to utilize during the addition activities.

I am adding ____ and ____.
First, I add the ____ place.
Next, I add the ____ place.
If there are 10 or more tens then ...
To regroup I ...
Finally, I add the numbers in the ____ place.
The sum is ____.

### English Language Development Leveled Activities

| Emerging Level | Expanding Level | Bridging Level |
|---|---|---|
| **Academic Word Knowledge** Write $142 + 73 =$ ____. Use base-ten blocks to model finding the sum, and have students follow along using their own manipulatives to model. After adding the tens, be sure to say, *I need to regroup.* Emphasize the regrouping of eleven tens into one hundred and one ten. Repeat with additional problems that require regrouping of tens. | **Choral Responses** Write $287 + 12 =$ ____. Model solving the problem, but pause when you need to add the tens column. Ask, *Regroup, or don't regroup?* Have students answer, **don't regroup.** Briefly discuss that the contraction *don't* means *do not.* Repeat with several more three-digit addition problems, randomly alternating between those that require regrouping of tens and those that do not. | **Public Speaking Norms** Write 162 and 83 on the board. Have students get into pairs, and say, *Write your own real-world, word problem that uses these two numbers and must be solved using addition. Then solve your problem.* Have students share their word problems and explain how they found their solutions. Be sure students use sentence frames such as: **There are 10 or more tens, so I must regroup 10 tens as 1 hundred.** |

### Multicultural Teacher Tip

Asian and Latin American students may avoid eye contact as you speak with them. This is easy to construe as shyness or disinterest, but extended eye contact may be viewed by these students as rude or as a challenge to your authority as the teacher. Similarly, a student may smile when he or she is being reprimanded or disagreeing with you, and the gesture might seem disrespectful. In fact, the student is most likely smiling to show respect in a difficult situation.

NAME _____ DATE _____

# Lesson 5 Concept Web
## *Regroup Tens to Add*

Use the concept web to show what you know about regrouping. Circle *yes* or *no*.

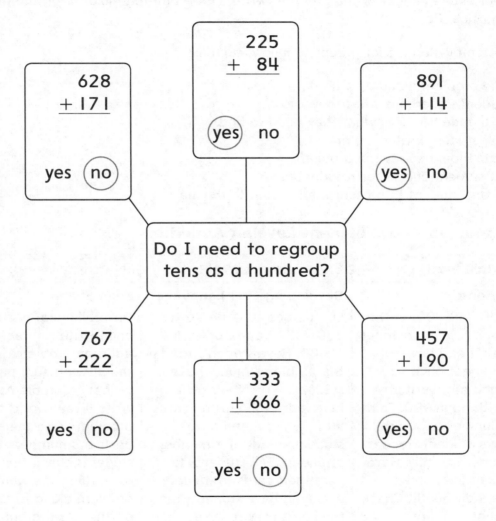

225
+ 84

(yes)　no

628
+171

yes　(no)

891
+114

(yes)　no

Do I need to regroup tens as a hundred?

767
+222

yes　(no)

333
+666

yes　(no)

457
+190

(yes)　no

**Teacher Directions:** Provide a description, explanation, or example of the term. Practice the sentence frame **[Yes/No], I [need/don't need] to regroup ten tens as one hundred.** After students have completed the web, have them orally compare answers with a partner.

# Lesson 6 Add Three-Digit Numbers

## English Learner Instructional Strategy

### Collaborative Support: Round the Table

Use a snipping tool to copy then paste each Problem Solving word problem in the lesson onto separate sheets of paper. Place students into multilingual groups of 4 or 5. Distribute one problem from Exercises 20–22 to each group. Have students work jointly on the problem by passing the paper around the table for each member to provide input. Direct each member of the group to write with a different color to ensure all students participate.

Provide a step-by-step list for groups to follow, such as:

1) Read the problem aloud as a group and discuss.
2) One student underlines what they know.
3) The next student circles what they need to find.
4) The next student writes a plan.
5) The next student solves the problem.
6) The last student checks for reasonableness.
7) Choose one student to present the solution to the class.

### English Language Development Leveled Activities

| Emerging Level | Expanding Level | Bridging Level |
|---|---|---|
| **Number Sense**<br>Write 1, 10, and 100 on the board. Point to each number as you say it and have students repeat each number chorally. Write by hundreds through 900. Point to each number as you say it and have students repeat each number chorally. Have students say the number sequence individually. Check that students are not adding the plural -s to each number. Repeat, counting backward by 100s chorally. | **Developing Oral Language**<br>On the board, write 200 + 300 = 500 and 800 + 100 = 900. Have students read the problems aloud. Have students write their own hundreds problems on cards. Sit in a circle and have students read their number problems. Tell students to pass their cards to the left, and then have students read the cards they receive. Continue moving cards around the circle until students read with reasonable fluency. | **Pairs Share**<br>Have bilingual partners each write a three-digit addition problem, including the answers. Tell partners to exchange problems and write a real-world story problem using their partner's numbers. Have students check their work by reading the word problem aloud to their partner. As a group, discuss the results. |

### Teacher Notes:

NAME _____   DATE _____

# Lesson 6 Guided Writing

## Add Three-Digit Numbers

**How do you add three-digit numbers?**

Use the exercises below to help you build on answering the Essential Question. Write the correct word from the word bank on the lines provided.

**1.** What key words do you see in the question?

add, three-digit, numbers

| **Word Bank** |
| :---: |
| hundred    ones    regroup    tens    three-digit number |

**2.** A digit is a symbol used to write numbers. The number 625 is a _three_-_digit_ _number_.

> Find 625 + 184.

**3.** Add the _ones_ first. If there are more than 10 ones, _regroup_ 10 ones as I ten.

**4.** Then add the _tens_. If there are more than 10 tens, regroup 10 tens as I _hundred_.

| hundreds | tens | ones |
| :---: | :---: | :---: |
| ☐ 1 | ☐ | |
| 6 | 2 | 5 |
| + 1 | 8 | 4 |
| 8 | 0 | 9 |

**5.** How do you add three-digit numbers?

First add the ones. Regroup. Add tens. Regroup. Add hundreds.

**Teacher Directions:** Read the Building on the Essential Question. Provide descriptions, explanations, or examples of the terms using images or real objects. Read each sentence frame and have students use the lesson and Glossary to write the appropriate terms. Have students read the sentences aloud.

**Grade 2 · Chapter 6** *Add Three-Digit Numbers* **55**

# Lesson 7 Rewrite Three-Digit Addition

## English Learner Instructional Strategy

### Graphic Support: Charts

Review a place-value chart that shows Tens and Ones. Remind students that place-value charts can be used to help line up vertical addition and subtraction problems. Draw a three-column place-value chart on the board. Ask students what to label the column to the left of tens. **hundreds** Have students copy the chart into their math journals.

Pair emerging/expanding students with bilingual (same native language) bridging level students during the Problem Solving exercises. Allow use of native language to clarify meaning. Encourage students to use manipulatives and a place-value chart as needed.

### English Language Development Leveled Activities

| Emerging Level | Expanding Level | Bridging Level |
|---|---|---|
| **Modeled Talk** | **Sentence Frames** | **Partners Work/Pairs Check** |
| Draw three columns on the board, and label them *Hundreds, Tens,* and *Ones.* Then write 132 + 175 = _____ horizontally on the board. For each number, point to each digit in turn as you ask, *hundreds, tens, or ones?* After students answer orally, record the digit in the correct column. Once all digits have been recorded in the columns, model vertical addition. Emphasize any regrouping that is necessary. Provide additional three-digit addition problems written horizontally. Have students rewrite problems vertically then add. | Write 453 + 278 = _____ horizontally on the board. Ask a volunteer to explain the first step in rewriting this problem as a vertical addition problem using the sentence frames: **The number _____ goes on top. The number _____ goes below it. Line up the numbers using _____. The sum goes _____.** Have volunteers take turns providing each step in guiding you to solve the problem, including explaining why you need to regroup the ones and tens. | Divide students into pairs. Have each pair write and solve an addition problem consisting of two three-digit numbers. Then have two sets of paired students work together to check answers. Each pair will take turns verbally instructing the other pair to solve their addition problem using base-ten blocks, including regrouping of ones and tens when necessary. |

## Teacher Notes:

NAME _____ DATE _____

# Lesson 7 Note Taking
## *Rewrite Three-Digit Addition*

Read the question. Write words you need help with. Use your lesson to write your Cornell notes.

| Building on the Essential Question | Notes: |
|---|---|
| **How can I rewrite three-digit addition?** | *Re-* means "again," so *rewrite* means "<u>write</u>  <u>again</u>." |

**Notes:**

*Re-* means "again," so *rewrite* means "<u>write</u>  <u>again</u>."

I can <u>rewrite</u> a problem to add.

> Find 333 + 279.

Write one addend <u>below</u> the other addend.

Line up the <u>ones</u>, <u>tens</u>, and <u>hundreds</u>.

$$\begin{array}{r} \boxed{1}\ \boxed{1}\phantom{0} \\ 3\ 3\ 3 \\ +\ 2\ 7\ 9 \\ \hline 6\ 1\ 2 \end{array}$$

Add.

If there are more than 10 <u>ones</u>, I should <u>regroup</u> as 1 <u>ten</u>.

If there are more than 10 <u>tens</u>, I should <u>regroup</u> as 1 <u>hundred</u>.

**Building on the Essential Question**

**How can I rewrite three-digit addition?**

**Words I need help with:**

See students' words.

**Teacher Directions:** Read the Building on the Essential Question and have students list words/phrases they need assistance with. Provide descriptions, explanations, or examples of the terms using images or real objects. Read each sentence frame and have students write the appropriate terms. Have students read their notes aloud.

# Lesson 8 Problem Solving
# Strategy: Guess, Check, and Revise

## English Learner Instructional Strategy

### Sensory Support: Manipulatives

Write *estimate*. Display a jar filled with items. Ask, *How many are there? Let's **guess**.* Write *guess* next to *estimate*. Point to each word as you say, *We can **guess** how many there are. Let's **estimate***. Write estimates. Say, *Let's check our guesses.* Count half of the objects so that students can judge half of the contents. Then ask, *Did we make good guesses? Do you want to revise your guess?* Write students' revised guesses. Finish counting the objects. Discuss results.

### English Language Development Leveled Activities

| Emerging Level | Expanding Level | Bridging Level |
|---|---|---|
| **Background Knowledge** | **Modeled Talk** | **Act It Out** |
| On the board, write: Understand, Plan, Solve, and Check. Write then say, *There are 125 eggs in a basket. 86 eggs are added to the basket. I guess there are 200 eggs. How many eggs are there?* Point to Understand on the board. Then ask, *Do we know how many eggs there are?* **no** Say, *I will circle what we need to find out.* Circle the last sentence. Point to Plan, and say, *Now we need a plan. We can Guess, Check, and Revise to solve.* Write *Guess, Check,* and *Revise* next to Plan on the board. Continue in this manner as you model solving the problem. | Have students sit in a circle. Show a container filled with marbles. Say, *I guess there are ____ marbles.* Pass the container to a student. Say, *I guessed that there are ____ marbles. What do you guess?* Have the students pass the container and give their estimates, using the sentence frame: **I guess there are ____ marbles.** Say, *Now we can check our guesses.* Remove about one-half of the marbles and count them. Then say, *____ is about half the number of marbles. I want to revise my guess. Now I guess ____.* Have students use the scaffold to revise their guesses. Count all the marbles and discuss. | Display a large container filled with many small items, such as beads, buttons, or marbles. Ask, *How can we estimate how many are in the jar if we don't count them?* Guide students to answer, **We can guess.** Have students make guesses about the number of objects. Then say, *If you want to check your guess, how would you do it?* Guide students to answer, **I would count some of the objects. Then I would revise my guess.** Count the number of objects in part of the container. Then ask students to revise their guesses and explain why they revised them. |

**Teacher Notes:**

NAME _____ DATE _____

# Lesson 8 Problem Solving
## STRATEGY: Guess, Check, and Revise

<u>Underline</u> what you know. (Circle) what you need to find. Guess, check, and revise to solve.

1. A store **sells** raisins in **packages** of 130, 165, and 170.

   **Rosa wants** to **buy** 300 raisins.

   Which two packages will **she** buy?

Raisins

| Guess: | Guess: | Guess: |
|--------|--------|--------|

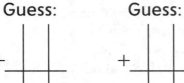

She will buy the packages of __130__ and __170__.

---

2. **Alexa's** ant **farm** has 120 ants.

   **She** buys **another** ant **farm**.

   She now has 300 ants.

   How many ants are in the **second** ant **farm**?

| Guess: | Guess: | Guess: |
|--------|--------|--------|
| 1  2  0 | 1  2  0 | 1  2  0 |
| +       | +       | +       |
|         |         | 3  0  0 |

There are __180__ ants in the second ant farm.

**Teacher Directions:** Provide a description, explanation, or example of the boldface terms and nouns using images or real objects. Read each sentence and have students echo read. Encourage students to guess, check, and revise to solve and then write their answers in each restated question. Have students read each answer sentence aloud.

**Grade 2 · Chapter 6** *Add Three-Digit Numbers* **57**

# Chapter 7 Subtract Three-Digit Numbers

## What's the Math in This Chapter?

### Mathematical Practice 8: Look for and express regularity in repeated reasoning

Distribute base-ten blocks to pairs of students and write the subtraction problem 856 − 321 vertically on the board. Using Virtual Manipulatives, have students follow along with their base-ten blocks as you model subtracting each place-value with base-ten blocks. Ask, *How many ones, tens, and hundreds are left after we have subtracted?* **5 hundreds, 3 tens, and 5 ones.** Record both the standard form and expanded form to make the connection between the manipulatives and the traditional algorithm.

Discuss with students how the subtraction process is similar to subtracting two-digit numbers. Model solving the problem using the traditional algorithm, placing it into a place-value chart if necessary. Be sure to make the connection between the answer that they found with their base-ten blocks to the traditional algorithm. Highlight that students can use the same algorithm repeatedly with any subtraction problem, beginning with the ones place and regrouping when needed.

Display a chart with Mathematical Practice 8. Restate Mathematical Practice 8 and have students assist in rewriting it as an "I can" statement, for example: **I can use repeated calculations I already know to solve a problem.** Post the new "I can" statement.

## Inquiry of the Essential Question:

### How can I subtract three-digit numbers?

Inquiry Activity Target: **Students come to a conclusion that they can use repeated calculations to subtract larger numbers.**

As an introduction to the chapter, present the Essential Question to students. The inquiry graphic organizer will offer opportunities for students to observe, make inferences, and apply prior knowledge of the subtraction process representing the Essential Question. As they investigate, encourage students to draw, write, and collaborate with peers to demonstrate their observations and thinking. Then have students present additional questions they may have to a peer to extend discussions.

Regroup students and restate Mathematical Practice 8 and the Essential Question. Pose questions to reflect on what has been learned to guide students in making connections between the Mathematical Practice and the Essential Question.

NAME _____ DATE _____

# Chapter 7 Subtract Three-Digit Numbers
## *Inquiry of the Essential Question:*

**How can I subtract three-digit numbers?**

Read the Essential Question. Describe your observations (I see...), inferences (I think...), and prior knowledge (I know...) of each math example. Write additional questions you have below.

**Mentally subtract 100.**

$$\begin{array}{r} 745 \\ -\ 100 \\ \hline \end{array}$$

THINK
$7 - 1 = 6$, so

$$\begin{array}{r} 745 \\ -\ 100 \\ \hline 645 \end{array}$$

I see ...

I think ...

I know ...

**Mentally subtract 10.**

$$\begin{array}{r} 831 \\ -\ 10 \\ \hline \end{array}$$

THINK
$3 - 1 = 2$, so

$$\begin{array}{r} 831 \\ -\ 10 \\ \hline 821 \end{array}$$

**502 − 98**

402  100

$100 - 98 = 2$

$402 + 2 = 404$

so, $502 - 98 = 404$

Take apart 502 as 402 and 100 since it is easier to subtract 98 from 100.

I see ...

I think ...

I know ...

Questions I have...

_____

_____

**Teacher Directions:** Read the Essential Question for students. Have students echo read. Direct students to describe their observations, inferences, and prior knowledge of each math example. Encourage students to write or draw additional questions they may have. Then have students share their ideas/questions with a peer.

# Lesson 1 Take Apart Hundreds to Subtract

## English Learner Instructional Strategy

### Sensory Support: Manipulatives

Before the lesson, write the terms: *ones, tens,* and *take apart* on a chart. Discuss the terms, write math examples, then model with a place-value chart and base-ten blocks to support understanding. Have students write the terms and draw a visual example in a journal. Next, have students write definitions and sentences using each term in context. Provide sentence examples if needed.

Post and model the following sentence frames to assist students during the lesson: **I will take apart _____ as _____ and 100. Then I will subtract _____ from 100 to get _____. I will add _____ back to _____ because it was taken from the 100. The difference is _____.**

## English Language Development Leveled Activities

| Emerging Level | Expanding Level | Bridging Level |
|---|---|---|
| **Choral Responses** | **Listen and Write** | **Public Speaking Norms** |
| Write several three-digit numbers with zeros on the board, such as 101, 200, and 806. Read the numbers aloud using the word *hundred*, for example: *one hundred one; two hundred; and eight hundred six.* Have students repeat chorally. Point to each digit in the numbers as you say it aloud, for example: *one, zero, one; two, zero, zero; eight, zero, six.* Have students repeat chorally. When you point to a zero, ask, *What is this called?* Have students answer, **zero.** | Describe a three-digit number containing zero(s) by saying each digit individually. For example, say, *three, three, zero.* Have students write each digit as you say it. Have students read the number aloud. For the given example, students should say, **three hundred thirty.** Repeat several more times with new three-digit numbers containing zero(s). | Read aloud a three-digit subtraction problem. Ask students to write it as you read. Then have students practice polite language as they seek clarification. For example, a student might ask, **Excuse me, could you repeat that please?** Have students work in pairs dictating problems to each other and asking politely for clarifications. |

**Teacher Notes:**

NAME _____ DATE _____

# Lesson I Note Taking
## *Take Apart Hundreds to Subtract*

Read the question. Write words you need help with and research each word. Use your lesson to write your Cornell notes. Write or draw math examples to explain your thinking. Share your examples with a classmate.

| **Building on the Essential Question** | **Notes:** |
|---|---|
| How can I take apart hundreds to subtract? | **Word Bank** — difference    take apart    subtract |

**Notes:**

**Word Bank**

difference    take apart    subtract

$$800 - 98 = ?$$

First, __take__ __apart__ 800 as 700 and 100.

Then __subtract__ 98 from 100.

The __difference__ is 2.

Add 2 to 700 to get the answer.

$800 - 98 = \underline{702}$.

**Words I need help with:**

See students' words.

**My Math Examples:**

See students' examples.

**Teacher Directions:** Read the Building on the Essential Question and have students list words/phrases they need assistance with. Provide descriptions, explanations, or examples of the terms using images or real objects. Read each sentence frame and have students write the appropriate terms. Have students read their notes aloud. Direct students to create a math example and share it with a peer.

Grade 2 • **Chapter 7** *Subtract Three-Digit Numbers* **59**

# Lesson 2 Subtract Hundreds

## English Learner Instructional Strategy

### Vocabulary Support: Activate Prior Knowledge

Use the Math Word Wall to review the terms: *subtraction, difference,* and *hundred.* Organize students into groups of 3–5 and assign a term to each group. Have students look up the term in the Glossary. Provide students with chart paper to write a definition, and draw a visual math example with the term labeled. Encourage use of native language to clarify meaning. Then, invite a volunteer from each group to share their definitions using the sentence frames: **Our math word is ____. It means ____.** Post the charts in the classroom, and have students record the terms and definitions in a journal for future reference.

### English Language Development Leveled Activities

| Emerging Level | Expanding Level | Bridging Level |
|---|---|---|
| **Making Connections** Write 400 − 300 = , then point to 4 and 3 and say, *4 take away 3 is 1, and 40 take away 30 is 10, so 400 take away 300 is 100.* Extend so as you speak. Write 500 − 300 = , then say, *5 minus 3 is 2, so 500 minus 300 is what?* Students may write, say, or gesture their answers. Say, *So the answer is . . .* to prompt the choral answer **200.** Repeat the activity as time allows, using similar subtraction problems. | **Word Meaning in Context** Draw a T-chart and label the headings with *away* and *a way.* Write 400 − 100 = ____. Say, *This problem is 400 take away 100,* as you point at *away.* Write then say, *Take away means "subtract."* Point at *a way.* Write then say, *We need a way to solve this problem. How can we solve this problem?* Stress the difference in sound between *away* and *a way.* Say, *A way to solve 400 take away 100 is to use base-ten blocks. Another way to solve 400 take away 100 is to use mental math. I know 4 take away 1 is 3, so 400 take away 100 is 300.* Write 500 − 200 = ____, and have students describe two ways to solve the problem. | **Exploring Language Structures** Write a subtraction problem on the board, for example: 300 − 200 = ____. Say, *I will take away 200 from 300 to find the answer.* Solve the problem. Say, *I took away 200 from 300 to get 100.* Emphasize the difference between *take* and *took.* Have students work in pairs. Each student will write a subtraction problem using multiples of one hundred and have his or her partner solve it. Have students ask, **What number did you take away from?** The partner answers, **I took away ____ from ____.** |

**Teacher Notes:**

NAME _____ DATE _____

# Lesson 2 Vocabulary Definition Map
## *Subtract Hundreds*

Use the definition map to write what the math word means and tell what the word is like. Write or draw a math example. Share your examples with a classmate.

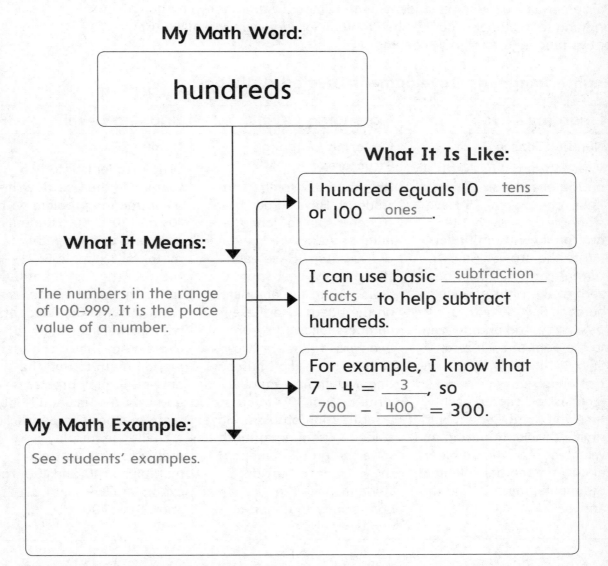

**My Math Word:**

hundreds

**What It Is Like:**

I hundred equals 10 __tens__ or 100 __ones__.

**What It Means:**

The numbers in the range of 100–999. It is the place value of a number.

I can use basic __subtraction__ __facts__ to help subtract hundreds.

For example, I know that 7 – 4 = __3__, so __700__ – __400__ = 300.

**My Math Example:**

See students' examples.

**Teacher Directions:** Provide a description, explanation, or example of the new term using images or real objects. Have students use the lesson or Glossary to define the math term. Direct students to list characteristics, and draw a picture representing their math term. Then encourage students to describe their picture to a peer.

**60   Grade 2 · Chapter 7** *Subtract Three-Digit Numbers*

# Lesson 3 Mentally Subtract 10 or 100

## English Learner Instructional Strategy

### Sensory Support: Pictures/Photographs

To introduce the lesson, use pictures of items in groups of 10 to practice counting as a class. Use pictures of the terms named throughout the lesson exercises to clarify the terms: *playground, flowers, bloomed, bugs, eaten, play* (theatrical production), *wildflowers, field, birds, pond, lake*, and *tadpole* to provide additional language support.

Pair emerging/expanding students with bilingual (same native language) bridging level students during the Problem Solving exercises. Allow use of native language to clarify meaning.

### English Language Development Leveled Activities

| Emerging Level | Expanding Level | Bridging Level |
|---|---|---|
| **Number Sense** | **Exploring Language Structures** | **Number Game** |
| Write $6 - 1 = $ ____ vertically on the board. Ask, *Six take away one is what?* **5** Have students answer chorally and/or show using fingers. Write the answer. Say, *The difference is five.* Add one zero to the right of each number. Say, *Sixty take away ten is fifty.* Add another zero to the right of each zero. Say, *Six hundred take away one hundred is five hundred.* Emphasize *hundred* each time you say it. Repeat with similar problems, having a volunteer read the problem aloud and another volunteer use mental math to find the answer. | Distribute 0–9 spinners to students. Have three students spin to generate a number. Ask, *What number did you spin?* Have each student answer, **I spun a ____.** Write the numbers on the board. Use these numbers followed by one or two zeros to create two-digit or three-digit numbers. Under each new number, write a 10 or 100 to create a subtraction problem. Have students use mental math to solve the problem. Repeat the activity so that different students have the opportunity to generate numbers. | Divide students into two teams. On the board, write a subtraction problem with 10s and 100s that students can solve using mental math, for example $569 - 10 = $ ____. The first team that solves the problem scores a point. Ask the first student who answers correctly to explain how he or she used mental math to find the answer. Display problems and award points until one team reaches five points. For a challenge, display problems that subtract multiples of 10 and 100. For example: $65 - 20 = $ ____, and $781 - 300 = $ ____. |

**Teacher Notes:**

NAME _____ DATE _____

## Lesson 3 Concept Web
### *Mentally Subtract 10 or 100*

Use the concept web to show different ways to subtract tens and hundreds.

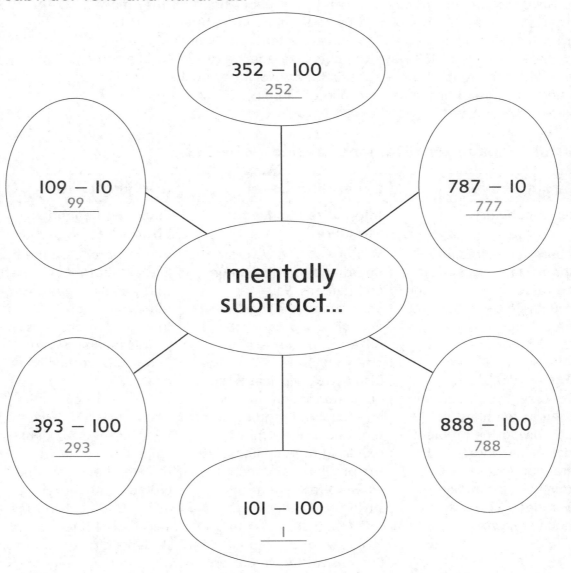

352 − 100
252

109 − 10
99

787 − 10
777

mentally subtract...

393 − 100
293

888 − 100
788

101 − 100
1

**Teacher Directions:** Review the terms *hundreds* and *tens*. The tell students that *mentally subtract* means to subtract in their heads. Before students begin the activity, have partners read the expressions to each other. Have partners complete the chart. Then have students switch partners and compare answers with another student using the sentence frame _____ **minus** _____ **equals** ____.

Grade 2 • **Chapter 7** *Subtract Three-Digit Numbers* **61**

# Lesson 4 Regroup Tens

*English Learner Instructional Strategy*

## Vocabulary Support: Cognate

Write *regroup* and its Spanish cognate, *reagrupar*, on a cognate chart. Assist non-Spanish speaking ELs with an appropriate translation tool.

Before the Explore and Explain page, give each student 123 straws or craft sticks. Have the students bundle the tens using rubber bands. If they make ten groups of 10, have them bundle those to make 100. Students should discover they all have a group of 1 hundred, 2 tens, and 3 ones. Ask, *How many straws are there?* **123** Have students take apart a group of tens and place it with the group of 3 ones. Discuss that everyone now has a group of 1 hundred, 1 group of ten and a group of 13 ones. Say, *We regrouped 1 ten. How many straws do we still have?* **123**

## English Language Development Leveled Activities

| Emerging Level | Expanding Level | Bridging Level |
|---|---|---|
| **Basic Vocabulary** | **Internalize Language Structure** | **Exploring Language Structures** |
| On the board, write 342 − 114. Use base-ten blocks and a hundreds, tens, and ones place-value chart to model regrouping of tens. Say, *I want to take away four ones. There are not enough ones to subtract from. I need to regroup. Say, regroup.* **regroup** Regroup one ten as ten ones to model solving the problem. Model similar three-digit subtraction problems regrouping tens, making sure students correctly pronounce *regroup*. | Write *do not = don't* on the board and briefly explain contractions. Say, *A contraction is a shortened way of saying two words together. In this example, do not can be said as don't.* Display two vertical three-digit subtraction problems, one requiring regrouping of tens and another not requiring regrouping. Solve the problems and identify when to regroup or not by saying, *I do need to regroup one ten as ten ones.* and *I don't need to regroup here.* Emphasize the word *don't*. Have students use the same phrases while discussing additional problems. | Write *regroup* on the board. Explain that re- is a prefix that means "again." Explain that when a prefix is added to a word, it changes the word's meaning. Write 245 − 682 vertically on the board. Say, *I did not write this problem correctly. I need to write it again. I need to rewrite it.* Put emphasis on the word *rewrite*, then X out the problem and write 682 − 245 next to the incorrect problem. Have students direct you to solve the problem using regrouping of tens. Repeat the activity, writing various three-digit subtraction problems. |

## Teacher Notes:

NAME _____ DATE _____

# Lesson 4 Vocabulary Sentence Frames
## *Regroup Tens*

The math words in the box are for the sentences below.
Write the words that fit in each sentence on the
blank lines.

| difference | regroup | subtract |
|---|---|---|

1. When I subtract, I am trying to find the __difference__.

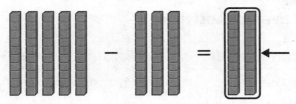

2. To __subtract__ is to take away one number from another
number.

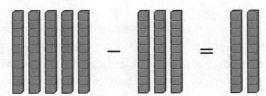

3. If I don't have enough ones, I can __regroup__ a ten as 10 ones.

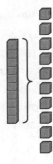

**Teacher Directions:** Provide a description, explanation, or example of the each term
using images or real objects. Read each sentence frame and have students echo read.
Direct students to write the correct terms in each blank. Then encourage students to
read each sentence to a peer.

# Lesson 5 Regroup Hundreds

## English Learner Instructional Strategy

### Collaborative Support: Mentor/Aide

Before the lesson, have a native language mentor/aide work with students to discuss subtracting three-digit numbers that require regrouping of hundreds. Have the mentor model and talk the students through the 3 steps shown in the See and Show box using Work Mat 7 and base-ten blocks.

To provide additional language support, review the following terms using photos/drawings to clarify meaning for the lesson/Problem Solving exercises: *macaroni, picture, basketball, camp, museum, party, parents,* and the season *fall.*

### English Language Development Leveled Activities

| Emerging Level | Expanding Level | Bridging Level |
|---|---|---|
| **Building Oral Language** | **Number Game** | **Partner Work** |
| On the board, write *Regroup* and *Do Not Regroup* as headings for a T-chart. Have students repeat the headings chorally. Write a three-digit subtraction problem in the Regroup column that requires regrouping of hundreds and another problem in the Do Not Regroup column that does not require regrouping of hundreds. Narrate solving each problem. Write another three-digit subtraction problem on the board. Have students use the T-chart examples to verbally guide you as you solve the problem. | Have students work in pairs. Each student will use number cubes to generate a three-digit number. Say to each pair of students, *Use your 2 three-digit numbers to create a subtraction problem.* If necessary, have each pair describe how they used regrouping to solve their problem. Provide the following sentence frames: **To regroup tens, I \_\_\_\_. To regroup hundreds, I \_\_\_\_.** | Write 518 − 263 = \_\_\_\_ vertically on the board. Say, *There were 518 fish in a lake. In one year, 263 fish were caught. How many fish are left?* Then model solving the problem after regrouping 1 hundred as 10 tens. Divide students into pairs. Write 435 − 190 = \_\_\_\_. Say, *Create your own real-world story that uses this problem and solve it.* Give pairs several minutes to come up with their stories and to solve the problem. Have students tell their stories and explain how they used regrouping to solve. |

**Teacher Notes:**

NAME _____ DATE _____

## Lesson 5 Vocabulary Word Study
### *Regroup Hundreds*

Circle the correct word to complete the sentence.

Regroup means to __take__ __apart__ a number to write it in a new way.

(take apart)          subtract

---

Show what you know about the word:

# regroup

There are __7__ letters.

There are __3__ vowels.

There are __4__ consonants.

__7__ letters in all – __3__ vowels = __4__ consonants.

---

Draw a picture to show what the word means.

See students' examples.

**Teacher Directions:** Provide a description, explanation, or example of the new term using base-ten blocks or real objects. Read the sentence and have students circle the correct word. Direct students to count the letters, vowels and consonants in the math term, then complete the subtraction number sentence. Guide students to draw a picture representing their math term. Then encourage students to describe their picture to a peer.

# Lesson 6 Subtract Three-Digit Numbers

## English Learner Instructional Strategy

### Vocabulary Support: Communication Guide

Post and model using the following communication guide for students to utilize during the subtraction activities.

I am subtracting the smaller number, _____, from the larger number, _____.

The numbers in the ones place are . . .

The numbers in the tens place are . . .

I cannot subtract _____ ones/tens from _____ ones/tens.

The numbers in the hundreds place are . . .

I will regroup because . . .

The difference is _____.

### English Language Development Leveled Activities

| Emerging Level | Expanding Level | Bridging Level |
|---|---|---|
| **Developing Oral Language** | **Word Meaning in Context** | **Understanding Oral Instructions** |
| Distribute base-ten blocks and Work Mat 7. Provide each pair with a different three-digit subtraction problem. Some problems should require regrouping and others should not. On the board, write the two following sentences: *We will regroup. We will not regroup.* Point to and model saying each sentence. Have students repeat chorally. Have pairs revisit their subtraction problems to decide which sentence correctly describes how it should be solved. Have students model with base-ten blocks, say the correct sentence, and then solve the problems. | Write 475 − 258 = _____ on the board. Say, *When I solve, I work from **right** to **left***. Regroup tens then solve the problem with the incorrect answer 119. Ask, *Is this the **right** answer?* **no;** The right answer is 219. Discuss the words *right* and *wrong,* and compare them to the words *left* and *right.* Write a new three-digit subtraction problem to the right of the previous one. Ask, *Is this new problem on the right or on the left of our first problem?* Prompt students to say, **The new problem is on the right.** Have students guide you in solving the new problem. | Write the following four directions on four individual index cards: 1) Write a problem that needs regrouping of tens. 2) Write a problem that needs no regrouping. 3) Write a problem that needs regrouping of both tens and hundreds. 4) Write a problem that needs regrouping only in the hundreds. Divide students into four teams. Give each team an index card, and have them write a subtraction problem. Each correct answer scores one point. Rotate index cards and repeat. After teams complete all four index cards, count points, and declare the winning team. |

**Teacher Notes:**

NAME _____ DATE _____

## Lesson 6 Flow Chart
### *Subtract Three-Digit Numbers*

Read the steps for subtracting three-digit numbers.
Put them in correct order in the flow chart.

**Steps for Subtracting**

Regroup I hundred as 10 tens.

Regroup I ten as 10 ones.

Subtract the hundreds.

Subtract the ones.

Subtract the tens.

| hundreds | tens | ones |
|:---:|:---:|:---:|
| 5 | II | II |
| 6̸ | 2̸ | X̸ |
| − 4 | 7 | 5 |
| 1 | 4 | 6 |

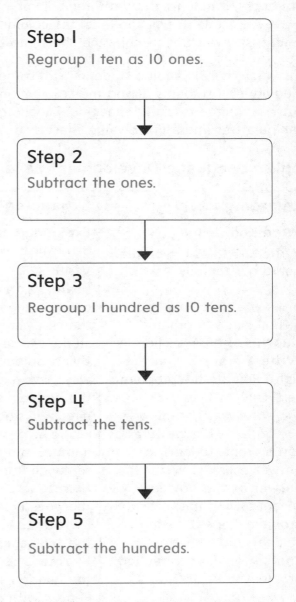

**Step 1**
Regroup I ten as 10 ones.

**Step 2**
Subtract the ones.

**Step 3**
Regroup I hundred as 10 tens.

**Step 4**
Subtract the tens.

**Step 5**
Subtract the hundreds.

**Teacher Directions:** Read the steps in the box. Review or clarify any unfamiliar terms. Have partners work together to write the steps in order on the flow chart. Then have students switch partners and describe the process to another student. Teach, model, and prompt sequence words such as *first, then, next, last,* and *finally*.

# Lesson 7 Rewrite Three-Digit Subtraction

## English Learner Instructional Strategy

### Graphic Support: Charts

Review a place-value chart that shows tens and ones. Remind students that the columns can be used to help line up vertical addition and subtraction problems. Draw a three-column place-value chart on the board. Ask students what to label the column to the left of the tens. **hundreds** Have students copy the chart into their math journals.

Pair Emerging/Expanding students with bilingual (same native language) Bridging level students during the Problem Solving exercises. Allow use of native language to clarify meaning. Encourage students to use manipulatives and a place-value chart as needed.

### English Language Development Leveled Activities

| Emerging Level | Expanding Level | Bridging Level |
|---|---|---|
| **Word Knowledge** Write $562 - 234 =$ on the board horizontally. Ask, *Do I write the answer on the left or on the right?* Use gestures as you prompt students to answer chorally: **Write the answer on the right.** Rewrite the problem vertically. Ask, *Do I write the answer at the top or at the bottom?* Use gestures as you prompt students to answer chorally: **Write the answer at the bottom.** Model solving the vertical problem. Repeat having students model with another similar subtraction problem. | **Exploring Language Structures** Write $325 - 187 =$ on the board. Ask a volunteer to come to the board and be the teacher. Prompt the student to say, **I will write this problem a different way.** Have the student rewrite the problem vertically. Then direct the student to describe where he or she placed each number using the following sentence frames: **I wrote ____ and ____ in the ones column. I wrote ____ and ____ in the tens column. I wrote ____ and ____ in the hundreds column.** Repeat with other students. | **Basic Vocabulary** Write 379, 126, 684. Have students create a subtraction problem with the largest and smallest three-digit number. **684 − 126 = ____** Write the subtraction problem both horizontally and vertically. Say, *These are two ways to write a subtraction problem.* Have students work in pairs to solve the problems both horizontally and vertically. When students have finished, lead a discussion about which method of solving was easier. Encourage students to use comparatives such as *easier* and *harder* in their responses. |

## Teacher Notes:

NAME _____ DATE _____

# Lesson 7 Note Taking

## *Rewrite Three-Digit Subtraction*

Read the question. Write words you need help with. Use your lesson to write your Cornell notes.

| **Building on the Essential Question** | **Notes:** |
|---|---|
| How do I rewrite two-digit subtraction? | *Re-* means "again," so *rewrite* means "__write__ __again__." |
| | I can __rewrite__ a problem to subtract. |
| | $$814 - 248 = ?$$ |
| | Write the __greater__ number on top. Write the other number __below__ it. |
| **Words I need help with:** | Line up the __ones__, tens, and __hundreds__. |
| See students' words. | $$\begin{array}{c c c} \boxed{7} & \boxed{10} & \boxed{14} \\ \cancel{8} & \cancel{1} & \cancel{4} \\ -\ 2 & 4 & 8 \\ \hline 5 & 6 & 6 \end{array}$$ |
| | Subtract. |
| | If there are not enough **ones**, I should __regroup__ I __ten__ as 10 ones. |
| | If there are not enough **tens**, regroup I __hundred__ as 10 tens. |

**Teacher Directions:** Read the Building on the Essential Question and have students list words/phrases they need assistance with. Provide descriptions, explanations, or examples of the terms using images or real objects. Read each sentence frame and have students write the appropriate terms. Have students read their notes aloud.

**Grade 2 · Chapter 7** *Subtract Three-Digit Numbers* **65**

# Lesson 8 Problem Solving Strategy: Write a Number Sentence

## *English Learner Instructional Strategy*

### Collaborative Support: Round the Table

Place students into multilingual groups of 4 or 5. Assign one problem from Exercises 1–6 to each group. Have students work jointly on the problem by passing the paper around the table for each member to provide input. Direct each member of the group to write with a different color to ensure all students participate. Provide a step-by-step list for groups to follow, such as: 1) Read the problem aloud and discuss. 2) One student underlines what they know and circles what they need to find. 3) The next student writes a plan. 4) The next student solves the problem and checks for reasonableness. 5) The last student presents the solution to the class.

### English Language Development Leveled Activities

| Emerging Level | Expanding Level | Bridging Level |
|---|---|---|
| **Build Background Knowledge** | **Building Oral Language** | **Public Speaking Norms** |
| On the board, write then read aloud, *325 cans were on a shelf. 154 cans were taken away. How many cans are left?* Ask, *How many cans were taken away?* Have students answer **154** chorally or with a gesture. Underline: *154 cans were taken away.* On the board, circle: *How many cans are left?* Say, *I underlined what we know. I circled what we need to find out.* Below the word Understand, write Plan, Solve, and Check. Model how to solve the problem and check the answer. Refer to each step written on the board as you complete it. | Write then read aloud: *A rancher has 550 horses. He wants to sell 75 horses. How many horses will he have left?* Write: Understand, Plan, Solve, and Check on four separate index cards with explanations for each step. Divide students into four groups and distribute one card to each group. Ask a volunteer from each group to read their card aloud, and then tell the group to direct you through the step. | Draw a large problem solving graphic organizer with four large boxes labeled: Understand, Plan, Solve, and Check. Have students work in pairs and distribute four index cards to each pair. Write then read aloud: *There are 365 days in a year. Today is day 186. How many days are left in the year?* Have pairs work together using the index cards for each step in the problem solving process. Circulate to answer any questions. Then have volunteers describe how they found the solution. Use the problem solving graphic organizer on the board to record their descriptions. |

**Teacher Notes:**

NAME _____ DATE _____

# Lesson 8 Problem Solving

## STRATEGY: Write a Number Sentence

<u>Underline</u> what you know. (Circle) what you need to find.
Write a number sentence to solve.

I. **It took** 227 days to build railroad
tracks over a **mountain**.

It took 132 days to build tracks
over **flat ground**.

How many more days **did it take**
to build over the mountain?

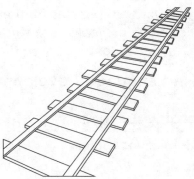

railroad tracks

227 ⊝ 132 ⊜ 95

It took __95__ more days to build over the mountain.

---

2. The **Fuller** family is driving 475 miles.

**They** have already **gone** 218 miles.

How many miles are left to go?

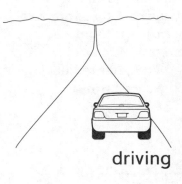

driving

475 ⊝ 218 ⊜ 257

They have __257__ miles left to go.

**Teacher Directions:** Provide a description, explanation, or example of the boldface
terms and nouns using images or real objects. Read each sentence and have students
echo read. Encourage students to write the subtraction number sentence and then
write their answers in each restated question. Have students read each answer
sentence aloud.

**66** Grade 2 • **Chapter 7** *Subtract Three-Digit Numbers*

# Lesson 9 Subtract Across Zeros

## English Learner Instructional Strategy

### Language Support: Echo Reading

Pair ELs with a native English-speaking student during Problem Solving Exercises 15 and 16. Have the native English-speaking student read the problem aloud, and have the English Learning student echo read. Then have students work together using a Four-Step Problem-Solving Plan online graphic organizer to solve each problem. Direct students to write a list of "clue" words or phrases from each problem that signaled them to write a subtraction sentence, such as: *leave, are still, are left,* and *how many more than.* At the end of the lesson, invite students to share their lists of "clue" words.

### English Language Development Leveled Activities

| Emerging Level | Expanding Level | Bridging Level |
|---|---|---|
| **Academic Vocabulary** | **Signal Words** | **Number Game** |
| Write $400 - 268 = $ ____ on the board. Distribute base-ten blocks and a hundreds, tens, and ones place value-chart to students and have them follow along as you model solving the problem. Involve students by asking simple questions that can be answered with one word or a gesture. For example, when you indicate there are zero ones or tens to take away from, ask, *Should we regroup?* After regrouping, ask, *Now how many?* | Write $600 - 364 = $ ____ on the board. Ask a volunteer to solve the problem. Afterward, have a second volunteer describe how the problem was solved. Provide the following sentence frames to help the student utilize signal words for order: **First he/she ____. Then he/she ____. Next he/she ____. Finally he/she ____.** Write a similar problem. Have a new pair of students solve the problem and describe the process. | Divide the class into three teams. Have one team come to the board while the other two teams turn away so they cannot see the board. Write $500 - 254 = $ ____. Have the team at the board solve the problem. When they are finished, have them erase one number from the problem. (Do not erase the answer.) Have the other teams turn around. The first team to correctly identify the missing number scores a point. Write new problems and have teams take turns solving and discovering the missing number. The first team to score four points wins. |

### Multicultural Teacher Tip

ELs may use an alternative algorithm when solving subtraction problems. In particular, Latin American students may have been taught the *equal additions method* of subtraction instead of the traditional US method of "borrowing" from the column to the left when the top number is less than the bottom number. In the *equal additions method,* a problem such as $35 - 18$ solved vertically would start with ten ones added to the top number ($15 - 8$) and then one ten is added to the bottom number ($30 - 20$), to get 7 and 10, or 17. Similarly, $432 - 158$ would be solved as $12 - 8$, $130 - 60$, and $400 - 200$ ($4 + 70 + 200 = 274$).

NAME _____ DATE _____

# Lesson 9 Flow Chart
## *Subtract Across Zeros*

Read the steps for subtracting across zeros. Fill in the missing words on the blank lines. Use your lesson to help.

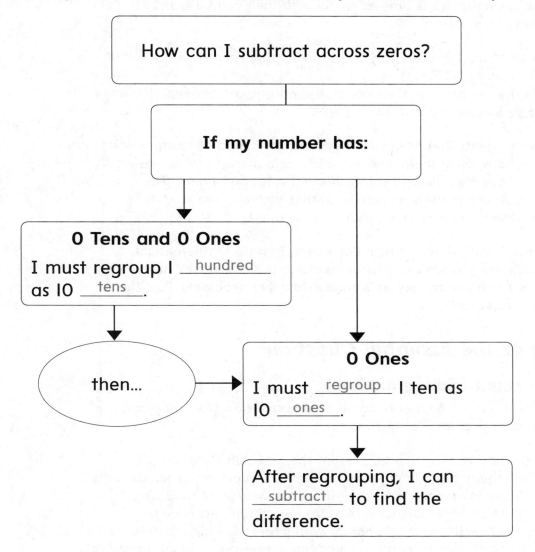

How can I subtract across zeros?

**If my number has:**

**0 Tens and 0 Ones**
I must regroup I ___hundred___
as I0 ___tens___ .

then...

**0 Ones**
I must ___regroup___ I ten as
I0 ___ones___ .

After regrouping, I can
___subtract___ to find the
difference.

**Teacher Directions:** Review the following terms as necessary: *ones, tens, hundreds, difference, regroup, subtract.* Model and prompt the If...then statements as necessary; for example, **If my number has _____, then _____.** Have partners work through the items together. Then have them switch partners and read their sentences.

Grade 2 • **Chapter 7** *Subtract Three-Digit Numbers* **67**

# Chapter 8 Money

## What's the Math in This Chapter?

### Mathematical Practice 4: Model with mathematics

Label an eraser 2¢, a pencil 5¢, a glue stick 13¢, and a box of markers 15¢. Distribute manipulative pennies, nickels, and dimes to each student. Display the eraser and say, *This eraser costs two cents. Show me two cents with your money.* Students should display two pennies. Repeat with the other three items, giving students time to model with their coins. Discuss all possible coin combinations.

Gesture to all of the items and say, *How much money would you need to buy all of the supplies? Please use your coins to **model** this problem.* Give students time to model with their coins offering support as needed. Discuss all possible coin combinations that equal 35¢.

Discuss with students that the coins they used are called concrete models because they can touch them. These models help them to solve real-world problems and see may ways to show different values of money. The discussion goal is that students recognize that they can use models to help them understand a problem that involves money.

Display a chart with Mathematical Practice 4. Restate Mathematical Practice 4 and have students assist in rewriting it as an "I can" statement, for example: **I can use money as a model to solve problems.** Post the new "I can" statement.

## Inquiry of the Essential Question:

### How do I count and use money?

Inquiry Activity Target: **Students come to a conclusion that they can use models to solve problems about money.**

As an introduction to the chapter, present the Essential Question to students. The inquiry graphic organizer will offer opportunities for students to observe, make inferences, and apply prior knowledge of modeling representing the Essential Question. As they investigate, encourage students to draw, write, and collaborate with peers to demonstrate their observations and thinking. Then have students present additional questions they may have to a peer to extend discussions.

Regroup students and restate Mathematical Practice 4 and the Essential Question. Pose questions to reflect on what has been learned to guide students in making connections between the Mathematical Practice and the Essential Question.

NAME _____  DATE _____

# Chapter 8 Money
## *Inquiry of the Essential Question:*

### How do I count and use money?

Read the Essential Question. Describe your observations
(I see...), inferences (I think...), and prior knowledge (I know...) of
each math example. Write additional questions you have below.

I see ...

I think ...

I know ...

(10¢)  (10¢)  (5¢)  (5¢)  (5¢)  (1¢)

__10__ ¢,  __20__ ¢,  __25__ ¢,  __30__ ¢,  __35__ ¢,  __36__ ¢ = __36__ ¢

---

I see ...

I think ...

I know ...

__25__ ¢,  __35__ ¢,  __45__ ¢,  __50__ ¢,  __51__ ¢,  __52__ ¢ = __52__ ¢

---

Questions I have...

_____

_____

 **Teacher Directions:** Read the Essential Question for students. Have students echo read.
Direct students to describe their observations, inferences, and prior knowledge of each
math example. Encourage students to write or draw additional questions they may have.
Then have students share their ideas/questions with a peer.

# Lesson 1 Pennies, Nickels, and Dimes

## English Learner Instructional Strategy

### Vocabulary Support: Math Word Wall

Add *penny*, *nickel*, and *dime* to the Math Word Wall. Add *cent*, *cents*, and *value* and their Spanish cognates, *centavo*, *centavos*, and *valor* to the cognate list. Organize students into groups of 3–5 students and assign a term to each group. Have students look up the term in the Glossary. Provide students with chart paper to write a definition, and draw a visual math example with the term labeled. Then, invite a volunteer from each group to share their definition using the sentence frame: **Our math word is _____. It means _____.** Post the charts in the classroom, and have students record the terms and definitions in their math journals for future reference.

### English Language Development Leveled Activities

| Emerging Level | Expanding Level | Bridging Level |
|---|---|---|
| **Basic Vocabulary** | **Money Sense** | **Internalize Vocabulary** |
| Distribute one penny, nickel, and dime to each student. Write each coin's name and their value on the board. Model saying them. Say, *A penny has a value of one cent. Show me a penny.* Students display pennies. Ask, *What is its value?* Prompt students to answer chorally, **one cent.** Have students look on the back of the penny to find the term *one cent.* Repeat with the other coins having students identify the terms *five cents* and *one dime* respectively on the back of the nickel and dime. | Write *penny*, *nickel*, and *dime* on index cards. Randomly distribute one card to each student. Invite a group of four students to the front of the classroom. Instruct students to stand in line in order of the value on their cards, greatest to least. Have each student say the coin on his or her card. Then ask, *What is the total value of the coins in line?* Have students work together to find the value and say it. For example: 10 cents + 5 cents + 5 cents + 1 cent = 21 cents. Repeat the activity with a new group of students. | Divide students into three groups. Give each group a different coin, either a penny, a nickel, or a dime and sticky notes. Say, *Use separate sticky notes to write four or more ways to describe your coin. Include the coin's value as one description.* Give groups several minutes to write their descriptions. Have students in each group take turns sharing one description with the class, then have them appropriately place the sticky notes on a three-column chart with the headings: Penny, Nickel, Dime. |

### Multicultural Teacher Tip

Because many word problems involve prices and/or determining changes in monetary value, ELs will benefit from an increased understanding of American coins and bills. A chart or other kind of graphic organizer visually comparing coin and bill values and modeling how to write dollars and cents in decimal form would help these students. You may also want to have ELs describe the monetary systems of their native countries. Identifying similarities or differences with the American system can help familiarize students with dollars and cents.

NAME _____ DATE _____

# Lesson I Word Identification
## *Pennies, Nickels, and Dimes*

Match each word to the pictures.

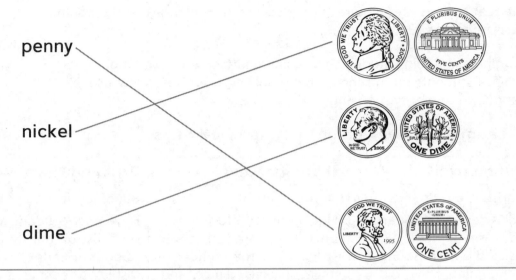

penny

nickel

dime

Write the correct word from above for each sentence
on the blank lines.

I. A __nickel__ is worth 5 cents.

2. A __dime__ is the smallest coin in size. It is worth 10 cents.

3. A __penny__ says "one cent" on the back.

**Teacher Directions:** Review the words using images or real objects. Have students say
each word then draw a line to match the word to its picture. Direct students to say each
word and then write the corresponding words in the sentences. Encourage students to
read the sentences to a peer.

**Grade 2 · Chapter 8** *Money* **69**

# Lesson 2 Quarters

## English Learner Instructional Strategy

### Sensory Support: Realia

To frontload basic and academic vocabulary before the lesson, use pictures, or realia (concrete objects) to review the following terms which may be unfamiliar to ELs: *penny, nickel, dime, quarter, sofa, enough, basketball, ticket, juice box, stuffed animal, toy, bracelet,* and *water*. Label the real objects and pictures and place prominently in the classroom for student reference.

Pair emerging/expanding students with bilingual (same native language) bridging level students during the Problem Solving exercises. Allow use of native language to clarify meaning.

### English Language Development Leveled Activities

| Emerging Level | Expanding Level | Bridging Level |
|---|---|---|
| **Non-Transferable Sounds** | **Number Game** | **Building Oral Language** |
| Write *quarter* on the board and model saying it. Emphasize the /kw/ sound at the beginning. Have students imitate your pronunciation. Say, *The value of one quarter is twenty-five cents. The value of two quarters is fifty cents.* Emphasize the /kw/ sound, and the plural /s/ at the end of *cents* versus the /z/ at the end of *quarters*. Repeat with the students identifying seventy-five cents and one-hundred cents, respectively. Throughout the activity, ask questions that prompt one-word or two-word answers, such as: **quarters; fifty cents.** | Divide students into two teams, Team A and Team B. Distribute a large number of pennies, nickels, dimes, and quarters to each team. Tell Team A to arrange enough pennies, nickels, and dimes to equal 25, 50, 75, or 100 cents. Then have a volunteer from Team A tell Team B how many of each coin they used, for example, **We used four dimes, four nickels, and fifteen pennies.** After discussion, prompt a volunteer from Team B to say how many quarters have the same value, for example, **Three quarters have the same value.** Have teams switch roles. | Have each student write a short list of four items for sale on an index card. Display a teacher example such as: eraser for 50¢, toy car for 100¢, pencil for 25¢, lollipop for 75¢. Say, *Price each item as either 25, 50, 75, or 100 cents.* Gather the cards and redistribute them to students along with four quarters apiece. Have students take turns saying what they would buy from their new list, and the number of quarters they need to buy it. For example, using the sentence frames: **I would buy the ____ for ____ cents. I would pay for it with ____.** |

### Teacher Notes:

NAME _____ DATE _____

# Lesson 2 Four-Square Vocabulary
## *Quarters*

Write the definition for *quarter*. Write what the word means, draw a picture, and write your own sentence using the word.

| **Definition** | **My Own Words** |
|---|---|
| quarter = 25¢ or 25 cents | See students' examples. |
| **My Picture** | **My Sentence** |
| See students' drawings. | Sample sentence: A quarter is bigger than a penny. |

**quarter**

 **Teacher Directions:** Provide a description, explanation, or example of the new term using images or real objects. Have students use the Glossary to write the definition. Direct students to write a definition in their own words and draw a picture representing their math term. Have students write a sentence using the term. Then encourage students to read their sentence to a peer.

# Lesson 3 Count Coins

## English Learner Instructional Strategy

### Vocabulary Support: Communication Guide

Post and model the following communication guide for students to utilize during the money activities.

The coins in the problem are . . .
The value of this coin is _____ cent(s).
A _____ is worth _____ cents.
The coin with the greatest value is _____.
First I skip counted by _____ then I skip counted by _____.
The value of all the coins is _____ cents.

### English Language Development Leveled Activities

| Emerging Level | Expanding Level | Bridging Level |
|---|---|---|
| **Money Sense**<br><br>Show students a combination of coins and model counting from greatest value to least value to find the total value. For example, show a quarter, three dimes, a nickel, and two pennies. Point to the appropriate coin and say, *25 cents, 35 cents, 45 cents, 55 cents, 60 cents, 61 cents, 62 cents. The value is 62 cents.* Show students other combinations of coins and have them count along with you to find the value. Provide the following sentence frame to help them state the total value of the group of coins: **The value is _____ cents.** | **Look, Listen, and Identify**<br><br>Distribute an index card to each student. Randomly assign each student a penny, nickel, dime, or quarter. Have students write the name of their assigned coin on the card. Choose five students to come to the front of the class. (Check that the value of their coins is at or under one dollar.) Have the remaining students direct them to stand in order from greatest value to least value as written on their cards. Ask, *What is the total value of the coins?* Have students count chorally as you point to each card to find the value. Repeat the activity. | **Exploring Language Structures**<br><br>Invite two students to the front of the classroom. Prompt each student to use the following sentence frame, _____ **(Mr./Ms. teacher name), please** *give* **me five coins.** Give each student a different combination of five coins. Then prompt them to say, _____ **(He/She)** *gave* **me five coins.** Emphasize the words *give* and *gave*. Ask students to count their coins and compare values. Have students work in groups of three to imitate the activity. After comparing values, students change roles and repeat the activity. |

**Teacher Notes:**

NAME _____ DATE _____

# Lesson 3 Note Taking
## *Count Coins*

Read the question. Write words you need help with. Use your lesson to write your Cornell notes. Write or draw math examples to explain your thinking. Share your examples with a classmate.

**Building on the Essential Question**

How can I count coins?

**Words I need help with:**

See students' words.

**Notes:**

A __penny__ is worth I cent (I¢).

A __nickel__ is worth 5 cents (5¢).

A __dime__ is worth I0 cents (I0¢).

A __quarter__ is worth 25 cents (25¢).

Start with the coin that has the __greatest__ value.

__Count__ to find the total.

I can also __skip__ __count__ by 5s and I0s to help me find the total.

**My Math Examples:**

See students' examples.

**Teacher Directions:** Read the Building on the Essential Question and have students list words/phrases they need assistance with. Provide descriptions, explanations, or examples of the terms using images or real objects. Read each sentence frame and have students write the appropriate terms. Have students read their notes aloud. Direct students to create a math example and share it with a peer.

Grade 2 • Chapter 8 *Money*  71

# Lesson 4 Problem Solving Strategy: Act It Out

## English Learner Instructional Strategy

### Collaborative Support: Act It Out

Before the lesson, use realia to review the following basic vocabulary which may be unfamiliar to students: *ring, piggy bank, toy truck, race car, cookie, gel pen,* and *notebook*. Label the real objects and place them around the classroom for students to reference as they complete the lesson. Pair students to collaboratively work on Exercises 1–6. Give each pair manipulative coins and have them act out counting and/or purchasing the items. When pairs have completed all 6 exercises, have them work with another pair to compare their solutions.

### English Language Development Leveled Activities

| Emerging Level | Expanding Level | Bridging Level |
|---|---|---|
| **Making Connections** | **Signal Words/Phrases** | **Order Words** |
| Draw a problem solving graphic organizer with four large boxes labeled: Understand, Plan, Solve, and Check. In the Understand box, draw an apple with a price tag showing 54¢. Display a quarter, two dimes, two nickels, and four pennies. Ask, *Do I have enough money for an apple?* In the Plan box, write: Act It Out. Say, *Our plan is to use money to act it out.* Model counting the coins then repeat counting, prompting students to count with you. Record the value of the coins in the Solve box. In the Check box, write: *Apple = 54¢, Coins = 59¢.* Model subtracting 54¢ from 59¢. Say, *Yes, I have enough money.* | Draw a problem solving graphic organizer with four large boxes labeled: Understand, Plan, Solve, and Check. Write then say, *A boy has two quarters, three dimes, one nickel, and three pennies. He spends 62¢. How much money does he have left?* Ask, *Do we need to add or subtract to find the solution?* **subtract** Guide a volunteer to draw a box around the words how much and left. Say, *These are signal words or clues that tell us to subtract.* Have students guide you in completing the other problem solving steps to find the solution. | Have students work in pairs. Distribute four index cards to each pair. Tell them to write: Understand, Plan, Solve, and Check on the cards. Write then say, *A girl has 67¢. She has one quarter, four nickels, and two pennies. The rest of her coins are dimes. How many dimes does she have?* Have pairs work together to find the answer using the problem solving steps written on the index cards. Ask volunteers to describe how they found the solution using order words: *first, second, next, then,* and *finally.* |

## Teacher Notes:

NAME _____ DATE _____

# Lesson 4 Problem Solving
## STRATEGY: Act It Out

<u>Underline</u> what you know. Circle what you need
to find. Act out each problem to solve.

I. **Maria** has I quarter in **her** piggy bank.

   Her mom gives her a nickel.

   Her dad gives her a dime.

   How much **money** does Maria have in all?

piggybank

She has ___40___ cents in all.

---

2. **Mark** has 2 quarters, I dime, and I penny.

   **He** wants to **buy** a toy truck for 55¢.

   Does he have **enough** money to buy the truck?

toy truck

___61¢___ > 55¢

___Yes___, he ___has___ enough money.

**Teacher Directions:** Provide a description, explanation, or example of the boldface
terms and nouns using images or real objects. Read each sentence and have students
echo read. Encourage students to act out the problem with coins and then write their
answers in each restated question. Have students read each answer sentence aloud.

# Lesson 5 Dollars

## English Learner Instructional Strategy

### Graphic Support: Utilize Resources

Write *dollar* and *dollar sign* and their Spanish cognates, *dólar* and *signo de dólar* on a cognate chart. Have students refer to the Glossary to write a definition and draw a visual math example in their math journals for future reference.

Review the difference between the dollar sign ($) and the cent sign (¢). Model and discuss how the dollar sign is written before the number and the cent sign is written after the number. Model how to properly read aloud $1 as, *one dollar* not *dollar one*.

### English Language Development Leveled Activities

| Emerging Level | Expanding Level | Bridging Level |
|---|---|---|
| **Number Recognition** | **Number Game** | **Public Speaking Norms** |
| Write and say, *one dollar.* Display a one dollar bill, and say, *This is one dollar.* Ask students to find the words *one dollar* on the bill, looking at both sides. To extend their thinking, ask students to count how many times the word *one* is on the one dollar bill. **8** Write: $1.00. Point to the $, and say, *This is the dollar sign.* Model how to write a dollar sign. Using write-on/wipe-off boards, have students practice writing dollar signs. | Create a set of ten flash cards. On four cards write: 100 pennies, 20 nickels, 10 dimes, and 4 quarters. On the other six cards, write amounts that do not equal one dollar, such as: 15 nickels, 25 dimes, 10 quarters, 50 pennies, 5 dimes, and 75 pennies. Say, *When I display a card that shows coins equal to one dollar, give me a thumbs-up and shout, "One dollar!" If the card does not show one dollar in value, say nothing and give me a thumbs-down.* Randomly display the cards one-by-one until students are recognizing those that show one dollar in value. | Have students work in pairs. Distribute a one dollar bill, a quarter, a dime, a nickel, a penny, and sticky notes to each pair. Have pairs discuss then write on the sticky notes how the coins and bill are alike (for example, they all have faces on them) and how they are different (each is worth a different amount). Using their sticky notes for reference, ask each pair to share their observations. |

**Teacher Notes:**

NAME _____ DATE _____

# Lesson 5 Vocabulary Definition Map
## *Dollars*

Use the definition map to write what the math word means and tell what the word is like. Write or draw a math example. Share your examples with a classmate.

**My Math Word:**

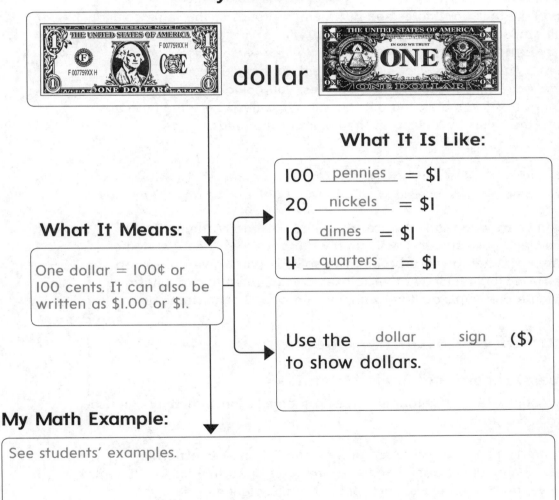

dollar

**What It Is Like:**

100 __pennies__ = $1

20 __nickels__ = $1

10 __dimes__ = $1

4 __quarters__ = $1

**What It Means:**

One dollar = 100¢ or 100 cents. It can also be written as $1.00 or $1.

Use the __dollar____sign__ ($) to show dollars.

**My Math Example:**

See students' examples.

**Teacher Directions:** Provide a description, explanation, or example of the new term using images or real objects. Have students use the lesson or Glossary to define the math term. Direct students to list characteristics, and draw a picture representing their math term. Then encourage students to describe their picture to a peer.

**Grade 2 · Chapter 8** *Money* **73**

# Chapter 9 Data Analysis

## What's the Math in This Chapter?

### Mathematical Practice 1: Make sense of problems and persevere in solving them

Write *red* on 3 sticky notes, *green* on 2 sticky notes, and *yellow* on 5 sticky notes. Display the 10 sticky notes scattered around the board. Say, *I took a survey of favorite colors. I need to know which color the most people like, but I don't know how to figure it out.*

Give students time to discuss with partners how they could organize the information. Ask, *What should I do first?* **Put the colors together in groups.** Take the sticky notes and put them in groups of the same color, but still randomly placed. Say, *Okay, I have the colors together, but I still can't tell which one has the most. I need to **persevere** in solving this problem. I need to organize this data or information. Could I move the sticky notes so they make more **sense**?* **Yes, you could line them up.** Take the groups of sticky notes and put them in vertical rows so students can compare the colors by looking at the height of the notes.

Say, *Now I can see that yellow is the favorite color of most of my friends. Which one is the least favorite?* **green** *How do you know?* **It is the shortest.** Say, *We persevered in solving the problem. We didn't give up.*

Display a chart with Mathematical Practice 1. Restate Mathematical Practice 1 and have students assist in rewriting it as an "I can" statement, for example: **When given a problem to solve, I don't give up. I keep trying.** Have students draw or write examples of when they did not give up solving a problem. Post the examples and new "I can" statement.

## Inquiry of the Essential Question:

### How can I record and analyze data?

Inquiry Activity Target: **Students come to a conclusion that they can use data to solve problems.**

As an introduction to the chapter, present the Essential Question to students. The inquiry graphic organizer will offer opportunities for students to observe, make inferences, and apply prior knowledge of data sets representing the Essential Question. As they investigate, encourage students to draw, write, and collaborate with peers to demonstrate their observations and thinking. Then have students present additional questions they may have to a peer to extend discussions.

Regroup students and restate Mathematical Practice 1 and the Essential Question. Pose questions to reflect on what has been learned to guide students in making connections between the Mathematical Practice and the Essential Question.

NAME _____  DATE _____

# Chapter 9 Data Analysis

## *Inquiry of the Essential Question:*

### How can I record and analyze data?

Read the Essential Question. Describe your observations (I see...),
inferences (I think...), and prior knowledge (I know...) of each math
example. Write additional questions you have below.

Quiz Scores

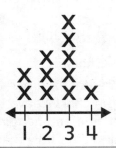

I see ...

I think ...

I know ...

| Favorite Theme Park Ride | | |
|---|---|---|
| Type of Ride | Tally | Total |
| Bumper Cars | \|\| | 2 |
| Ferris Wheel | \|\|\| | 3 |
| Roller Coaster | ⵏ \|\| | 7 |

I see ...

I think ...

I know ...

| Type of Ride | | | | | | |
|---|---|---|---|---|---|---|
| Bumper Car | ▨ | ▨ | | | | |
| Ferris Wheel | ▨ | ▨ | ▨ | | | |
| Roller Coaster | ▨ | ▨ | ▨ | ▨ | ▨ | ▨ |

Key: Each ▨ = I vote

I see ...

I think ...

I know ...

## Questions I have...

_____

_____

**Teacher Directions:** Read the Essential Question for students. Have students echo read
Direct students to describe their observations, inferences, and prior knowledge of each
math example. Encourage students to write or draw additional questions they may have.
Then have students share their ideas/questions with a peer.

# Lesson 1 Take a Survey

## English Learner Instructional Strategy

### Vocabulary Support: Frontload Academic Vocabulary

Before the lesson, write *survey, tally marks, and data* (*datos,* Spanish cognate) on the Math Word Wall. Introduce the words, write a math example, and provide concrete objects to support understanding. Organize students into groups of 3–5 and assign a term to each group. Have students look up the term in the glossary. Provide students with chart paper to write a definition, and draw a visual math example with the term labeled. Then, invite a volunteer from each group to share their definitions using the sentence frames: **Our math word is ____. It means ____.** Post the charts in the classroom, and have students record the terms and definitions in their math journals for future reference.

### English Language Development Leveled Activities

| Emerging Level | Expanding Level | Bridging Level |
|---|---|---|
| **Building Oral Language**<br><br>Draw a two-column chart with the headings, Boys and Girls. Say, *I will ask a question. I will take a survey.* Write then say, *Are there more boys or girls in our class? Let's take a survey to find out.* Have students count the male students. As they count, make tally marks on the board under Boys. Then have students count the female students while you make tally marks under Girls. Circle the tally marks, then say, *This is data.* Model how to count the tally marks for both the boys and girls. Repeat having students count chorally. Ask, *Are there more boys or girls in our class?* Have students respond according to data results. | **Exploring Language Structures**<br><br>Say, *I will take a survey. I will collect data using tally marks. Do you have a pet? Raise your hand if your answer is yes.* Have a student count the raised hands while you make tally marks on a two-column chart. Repeat, asking for no answers. As you record the tally marks, ask, *What am I doing?* Prompt students to answer chorally, **You are taking a survey.** Ask, *How did I take a survey?* Have students answer, **You asked a question. You used tally marks to show the data.** Ensure students use inflectional endings -ing and -ed correctly. | **Pairs Share**<br><br>On the board, draw a three-column chart. Leave the chart title and headings blank. Write tally marks to show 8, 12, and 3 under the blank headings. Say, *This is the data from a survey. There is no title and no headings.* Have students work in pairs and create their own survey question which supports the data. Direct students to draw a chart and write an appropriate title and headings. Have pairs share their survey using complete sentences. For example, **Our question is: Which animal is the best pet? Our title is: Best Pet. 8 students answered dog, 12 students answered cat, and 3 students answered fish.** |

## Multicultural Teacher Tip

Some ELs may write tally marks in a somewhat different format than the familiar four vertical lines with one diagonal. In Latin America, tally marks are generally drawn as four successive lines that form a square, with the fifth mark being a diagonal across the square. Students from Asia will use marks that eventually form the following character consisting of five marks: 正

NAME _____ DATE _____

# Lesson I Note Taking
## *Take a Survey*

Read the question. Write words you need help with.
Use your lesson to write your Cornell notes. Write or
draw math examples to explain your thinking. Share
your examples with a classmate.

| | |
|---|---|
| **Building on the Essential Question**<br><br>How can I take a survey? | **Notes:**<br><br>When I take a survey, I ask a ___question___ .<br><br>I know that I can use ___tally___ ___marks___ to record the answers, or ___data___ .<br><br>I know that data is ___information___ .<br><br>First, I write a ___question___ .<br><br>Then, I ___ask___ my question.<br><br>Next, I ___record___ each person's ___answer___ with a tally mark.<br><br>Finally, I ___count___ the tally marks. |
| **Words I need help with:**<br>See students' words. | |

**My Math Examples:**
See students' examples.

**Teacher Directions:** Read the Building on the Essential Question and have students list
words/phrases they need assistance with. Provide descriptions, explanations, or examples
of the terms using images or real objects. Read each sentence frame and have students
write the appropriate terms. Have students read their notes aloud. Direct students to
create a math example and share it with a peer.

**Grade 2 • Chapter 9** *Data Analysis* **75**

# Lesson 2 Make Picture Graphs
## English Learner Instructional Strategy

### Vocabulary Support: Math Word Wall

Before the lesson, write *picture graph* and *symbol* on the Math Word Wall. Introduce the words, write a math example, and provide concrete objects to support understanding.

English learners may need practice pronouncing the /f/ sound in *graph*. Help them generate a list of words with the /f/ sound that highlights the various spellings: f, ff, -fe, ph-, and -gh. Such as: *fall, February, giraffe, stuff, off, life, wife, telephone, dolphin,* and *laugh.*

### English Language Development Leveled Activities

| Emerging Level | Expanding Level | Bridging Level |
|---|---|---|
| **Word Recognition** | **Background Knowledge** | **Sentence Frames** |
| Conduct a survey of favorite fruits. Use data to create a picture graph. Say, *This is a picture graph.* Have students repeat chorally. Label the graph. Write then say, *1 piece of fruit stands for 1 vote.* Point to one symbol in the picture graph. Ask, *How many votes?* Students answer chorally, **one vote.** Prompt students to count symbols in one category. Ask, *How many votes in all?* Students answer chorally or with gestures. | Ask, *How many students are boys?* Use tally marks to record the answer. Repeat for girls. Say, *We will make a picture graph. What symbol will we use for boys?* Draw two or three student suggestions on the board. Survey the students to see which symbol is liked the best. Have a volunteer use tally marks to record their votes. Repeat to determine a symbol for girls. Have students create a picture graph with the original survey data. Then have students work with a partner to compare and contrast their graphs. | Display a blank picture graph. Say, *I want to take a survey of students' favorite zoo animal. Then I want to use this picture graph to show the results.* Have students direct you in taking the survey. Use the data to make a picture graph. Provide the following sentence frames: **The question we will ask is _____. The symbol for _____ will be _____.** |

**Teacher Notes:**

NAME _____ DATE _____

# Lesson 2 Definition Map
## *Make Picture Graphs*

Use the definition map to write a description and list characteristics about the vocabulary word or phrase. Write or draw math examples. Share your examples with a classmate.

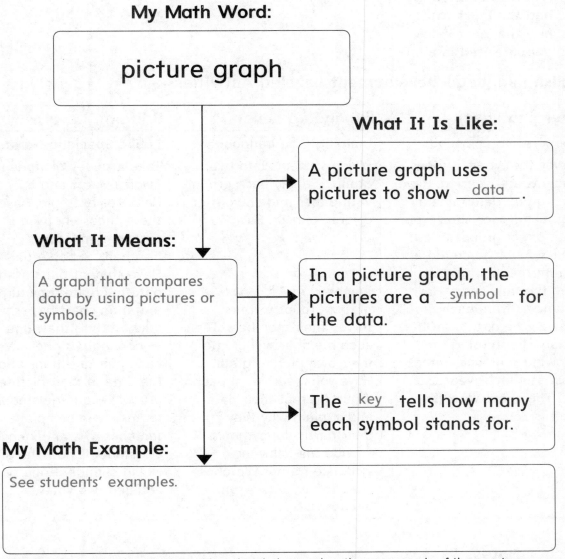

**My Math Word:**

picture graph

**What It Is Like:**

A picture graph uses pictures to show ___data___.

**What It Means:**

A graph that compares data by using pictures or symbols.

In a picture graph, the pictures are a ___symbol___ for the data.

The ___key___ tells how many each symbol stands for.

**My Math Example:**

See students' examples.

**Teacher Directions:** Provide a description, explanation, or example of the new term using images or real objects. Have students use the lesson or Glossary to define the math term. Direct students to list characteristics, and draw a picture representing their math term. Then encourage students to describe their picture to a peer.

# Lesson 3 Analyze Picture Graphs

## English Learner Instructional Strategy

### Language Structure Support: Communication Guide

Post and model how to use the following communication guide. Have students utilize the sentence frames during the picture graph activities.

The survey is about _____.

The key shows that each _____ stands for _____.

The symbol for _____ is _____.

_____ received _____ votes.

_____ had the most votes.

_____ had the least votes.

_____ people voted in all.

### English Language Development Leveled Activities

| Emerging Level | Expanding Level | Bridging Level |
|---|---|---|
| **Cooperative Learning**<br><br>Survey the types of shoes students are wearing. Have each group determine a symbol for each shoe type. Then direct groups to use their symbols to create a picture graph with data from the tally chart. Have volunteers from each group analyze the data from the picture graph using the following sentence frames: **Most students wear _____,** and **The shoe worn the least is _____.** | **Building Oral Language**<br><br>Divide students into two groups. Supply each group with paper and crayons or colored pencils. Say, *Each group will take a survey of their favorite season. Choose a symbol for each season, and then make a picture graph.* Groups will first survey their members. Then each member will create their own picture graph. Have volunteers from each group present their data and explain why they chose their particular symbols. Provide the following sentence frame: **We chose _____ for _____ (season) because _____.** | **Public Speaking Norms**<br><br>Take a survey of students' favorite sport and tally the data. For example, have them choose between soccer, gymnastics, baseball, or basketball. Have student pairs use the data to create a picture graph on chart paper with a key stating that: *one symbol equals one vote.* Ask each pair questions about the data in their picture graph. Encourage students to answer in complete sentences. As an extension, ask students how the graph would change if one symbol equaled two votes. |

**Teacher Notes:**

NAME _____ DATE _____

# Lesson 3 Word Identification
## *Analyze Picture Graphs*

Match each word to the correct part of the graph.

picture graph

symbol

key

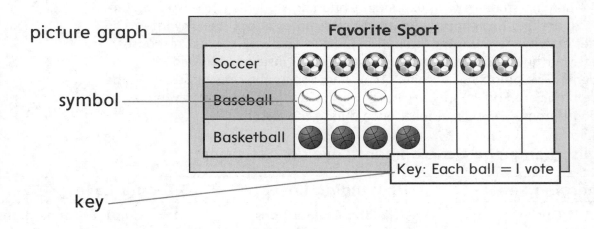

**Favorite Sport**

| Soccer | | | | | | | | |
| Baseball | | | | | | | | |
| Basketball | | | | | | | | |

Key: Each ball = I vote

Write the correct term from above for each sentence.

1. The __key__ tells how many each symbol stands for.

2. The __picture__ __graph__ tells our favorite sports.

3. Each __symbol__ stands for data.

**Teacher Directions:** Review the terms using images or real objects. Have students say each term and then draw a line to match the term to the corresponding part of the picture graph. Direct students say each sentence and then write the appropriate words in the sentences. Encourage students to read the sentences to a peer.

Grade 2 • **Chapter 9** *Data Analysis* **77**

# Lesson 4 Make Bar Graphs
## English Learner Instructional Strategy

### Collaborative Support: Think-Pair-Share

Before the lesson, write *bar graph* and its Spanish cognate, *gráfica de barros,* on the cognate chart. Introduce the word, write a math example, and provide concrete objects to support understanding.

Pair emerging students with expanding/bridging students while working on the Explore and Explain page to ask 10 classmates about their favorite winter activity. As you work through the lesson and seek student responses, direct your questions or prompts to student pairs instead of individual students. Give pairs time to think about and discuss their response. Allow the more proficient English speaker to answer for the pair. Be sure to prompt a response from each pair at least once during the lesson.

### English Language Development Leveled Activities

| Emerging Level | Expanding Level | Bridging Level |
|---|---|---|
| **Act It Out** | **Making Connections** | **Make Cultural Connections** |
| Place a stack of construction paper with various colors on a table. Have each student select a sheet. Direct students to group themselves by color. Prompt students to say, **We will make a bar graph.** Direct students in each group to line their papers up on the floor end-to-end. Use sentence strips to write a title and labels. Discuss how bar graphs are different from tally charts and picture graphs. | Display a tally chart from a previous lesson. Demonstrate how to use the data to make a bar graph. Discuss how bar graphs are different from tally charts and picture graphs using the sentence frames: **Bar graphs are like picture graphs because ___. They are different because ___.** Post a different tally chart. Say, *Use this tally chart to make a bar graph.* Have students work in pairs to create a bar graph using the tally chart data. Guide pairs to properly title and label their bar graphs. | Have students conduct a survey about which countries their classmates were born in. Monitor their progress and provide necessary help as they write their questions and survey their classmates. After students convert their survey into a bar graph, ask them to report their data using complete sentences. Provide the following sentence frames: **I surveyed ___ students. ___ students were born in ___ (country name).** |

### Teacher Notes:

NAME _____ DATE _____

# Lesson 4 Guided Writing
## *Make Bar Graphs*

**How do you make a bar graph?**

Use the exercises below to help you build on answering the Essential Question. Write the correct word or phrase on the lines provided.

1. **What key words do you see in the question?**
   bar graph

2. **What is it called when you collect information by asking people the same question?** survey

3. **What are numbers collected from a survey called?** data

4. **You can use data in a** tally chart **to make a bar graph.**

5. **On a bar graph, color one box for each piece of** data **.**

6. **How do you make a bar graph?**

   Do a survey and collect data in a tally chart. Color one box on the bar

   graph for each vote.

**Teacher Directions:** Read the Building on the Essential Question. Provide descriptions, explanations, or examples of the terms using images or real objects. Read each sentence frame and have students use the lesson and Glossary to write the appropriate terms. Have students read the sentences aloud.

# Lesson 5 Analyze Bar Graphs
## English Learner Instructional Strategy

### Language Structure Support: Report Back

Before the On My Own exercises, model how to restate the question in the answer. For example, write then read aloud Exercise 4, *What color hair do 5 students have?* Write then say, *Five students have _____ hair.* Direct students to look at the bar graph for the answer. Then prompt students to use the restated sentence frame to chorally respond: **Five students have brown hair.** As students work independently on Exercises 5–13, periodically ask students to individually report back to you restating the question in their answer.

### English Language Development Leveled Activities

| Emerging Level | Expanding Level | Bridging Level |
|---|---|---|
| **Choral Responses** | **Exploring Language Structure** | **Internalizing Vocabulary** |
| Survey students to determine their favorite ice cream flavor. Display pictures showing ice cream. Ask, *Is your favorite ice cream flavor vanilla, chocolate, or strawberry?* Record the data in a bar graph. Point to the shortest bar. Ask, *Does this show the most or the fewest?* **the fewest** Point to the tallest bar, and ask, *Does this show the most or the fewest?* **the most** Model comparing other bars in the graph using the terms more and less. For example, *Do more students like strawberry ice cream or chocolate ice cream?* **Sample answer: chocolate** | Have students work in pairs. Distribute a bag containing four colored blocks to each pair. Say, *You will draw colored blocks from the bag. Then you will make a bar graph to show the colors you drew from the bag.* Emphasize the difference between *draw* and *drew*. Have students draw blocks from the bag and tally the color they draw each time. Direct students to draw ten times. Then have students use their data to create a bar graph to show how many times each color was drawn. Have each pair describe their data using the word *drew*. | Have students randomly take pattern blocks from a bag and record the shapes they draw each time on a tally chart. Make sure students return drawn blocks to the bag before they draw again. Direct students to draw blocks ten times. Then have students create a bar graph using data from their tally chart to show how many times each shape was drawn. Have students analyze the data in their bar graphs using the comparatives and superlatives: *more, most, fewer,* and *fewest.* |

**Teacher Notes:**

NAME _____ DATE _____

# Lesson 5 Vocabulary Sentence Frames
## *Analyze Bar Graphs*

The math words in the box are for the sentences below. Write the words that fit in each sentence on the blank lines.

| survey | bar graph | data |
|--------|-----------|------|

**I.** Numbers or symbols that show information is __data__ .

| Name | Number of Brothers |
|------|--------------------|
| Katie | 2 |
| Lila | 0 |
| Blair | I |

**2.** Data can be shown in a __bar__ __graph__ .

**Number of Brothers**

| Katie | | | | | | |
|-------|--|--|--|--|--|--|
| Lila | | | | | | |
| Blair | | | | | | |

**3.** The __survey__ question was, "How many brothers do you have?"

**Teacher Directions:** Provide a description, explanation, or example of the each term using images or real objects. Read each sentence frame and have students echo read. Direct students to write the correct terms in each blank. Then encourage students to read each sentence to a peer.

Grade 2 • **Chapter 9** *Data Analysis* **79**

# Lesson 6 Problem Solving Strategy: Make a Table

## English Learner Instructional Strategy

### Collaborative Support: Echo Reading

Pair ELs with a native English-speaking student during the Problem Solving exercises. Have the native English-speaking student read the problem aloud, and have the EL echo read. Then have students work together to complete the tables for each exercise.

### English Language Development Leveled Activities

| Emerging Level | Expanding Level | Bridging Level |
|---|---|---|
| **Word Knowledge** | **Building Oral Language** | **Multiple Meaning Word** |
| Display a toy car or picture of a car and ask, *What is this?* Students answer chorally, **a car.** Model and say, *One car has four wheels. How many wheels on three cars?* Draw a table with two columns and five rows. Label the column headings, Cars and Wheels. Have students draw the same table. Point to the table, and say, *This is a table.* Have students repeat chorally. Use toy cars or pictures as you model completing the table. Have students complete their tables. Then discuss how making a table helps to solve a problem. | Write then say, *Five cars can be washed in an hour. How many cars can be washed in six hours?* Draw a table with two columns and six rows. Label the column headings, Cars and Hours. Using the problem solving steps; Understand, Plan, Solve and Check, guide students step-by-step as they solve the problem. Provide the following sentence frames: **I know that _____. I need to find out _____. I can _____ to solve the problem. The solution is _____.** | Write then say, *A squirrel eats three acorns each day. How many acorns will the squirrel eat in four days?* Draw a table that can be used to solve the problem. Say, *This is a table. We can use the table to find the answer.* Explain that some words have more than one meaning. Ask, *What is another meaning for table?* Guide students to respond referencing a table used for eating. Have students work in pairs. Distribute a problem solving graphic organizer to each pair, and tell students to complete it as they solve the problem using the table. |

**Teacher Notes:**

NAME _____ DATE _____

# Lesson 6 Problem Solving
## STRATEGY: Make a Table

Underline what you know. Circle what you need to find. Make a table to solve each problem.

**1. Desiree** has **4 pairs** of socks in **her drawer.**

How many socks are there in all?

pair of socks

| Pair | Socks in All |
|------|--------------|
| 1 | 2 |
| | |
| | |
| | |

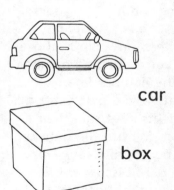

drawer

There are __8__ socks in all.

---

**2. Mr. Minnick** needs to **deliver** 60 boxes. **His** car can **hold** 10 boxes at a time.

How many **trips will he** need to **make** to deliver all 60 boxes?

car

box

He will need to make __6__ trips.

**Teacher Directions:** Provide a description, explanation, or example of the bold face terms and nouns using images or real objects. Read each sentence and have students echo read. Encourage students to make a table to solve the problem and then write their answers in each restated question. Have students read each answer sentence aloud.

# Lesson 7 Make Line Plots

## English Learner Instructional Strategy

### Vocabulary Support: Word Webs

Write *line* and its Spanish cognate, *línea* on a cognate chart. Review the math terms *line* and *number line* and display visual examples to support understanding.

Write line at the center of a word web on chart paper. Ask students to share all their non-math connections to the term *line* and add them to the web such as: *lunch line, lines on parking spaces and on roads, lines in pictures, lines used on sport fields/courts, lines on clothing,* etc. Add the term *line plot* to the web then begin the Explore and Explain page.

### English Language Development Leveled Activities

| Emerging Level | Expanding Level | Bridging Level |
|---|---|---|
| **Sentence Frames** <br><br> Draw a line on the board. Say, *This is a line.* Draw ten increments on the line and write the numbers 0–9. Say, *This is a number line.* Ask each student, *How old are you?* Have students answer using the following sentence frame: **I am _____ years old.** As each student answers, draw an X above the corresponding number on the number line to create a line plot. Once all students have said their ages, point to the data on the board, and say, *This is a line plot.* Discuss with students that each X represents the age of one student. | **Exploring Language Structure** <br><br> Distribute a number cube to each student and one for yourself. Say, *I will **roll** my number cube. I **rolled** a _____.* Emphasize the difference between *roll* and *rolled.* Draw a line plot, and record the number you rolled by writing an X above the number. Have students roll their number cubes, and then ask each student, *What did you roll?* Have the student answer, **I rolled a _____.** Record students' answers on the line plot. Have students first discuss with a partner what the X's mean, then discuss as a class. | **Public Speaking Norms** <br><br> Create a tally chart to show the number of cups of milk students drink daily. Ask each student how many cups of milk they drank yesterday, then have students record their responses on the tally chart. Have students guide you in using the data to create a line plot. Encourage students to speak in complete sentences using the terms: *tally mark(s), tally chart, line plot, X(s),* and *data.* |

### Teacher Notes:

NAME _____ DATE _____

# Lesson 7 Note Taking
## *Make Line Plots*

Read the question. Write words you need help with and research each word. Use your lesson to write your Cornell notes.

| **Building on the Essential Question** | **Notes:** |
|---|---|

**Building on the Essential Question**

How can I make line plots?

**Words I need help with:**

See students' words.

**Notes:**

A line plot is a way to ___organize___ data.

It shows how often a certain number ___occurs___ in data.

| Number of Cats | |
|:---:|:---:|
| 0 | &#124;&#124; |
| 1 | (&#124;&#124;&#124;) ← |
| 2 | |
| 3 | &#124; |
| 4 | &#124; |
| 5 | |

The tally marks are called ___data___.

Draw an X above the ___number___ for each piece of data.

Number of Cats

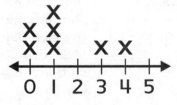

**Teacher Directions:** Read the Building on the Essential Question and have students list words/phrases they need assistance with. Provide descriptions, explanations, or examples of the terms using images or real objects. Read each sentence frame and have students write the appropriate terms. Have students read their notes aloud.

Grade 2 • **Chapter 9** *Data Analysis* **81**

# Lesson 8 Analyze Line Plots
## English Learner Instructional Strategy

### Language Structure Support: Report Back

During the See and Show exercises, model how to restate the question in the answer. For example, write then read aloud Exercise 1, *How many students have 2 hats?* Write then say, _____ *students have 2 hats.* Direct students to look at the line plot for the answer. Then prompt students to use the restated sentence frame to chorally respond: **Six students have 2 hats.** Repeat with Exercises 2–4. As students work through the On My Own exercises, periodically ask students to individually report back to you restating the question in their answer.

### English Language Development Leveled Activities

| Emerging Level | Expanding Level | Bridging Level |
|---|---|---|
| **Build Background Knowledge** Have students write their first names, and then have them count the number of letters used to spell their name. Record the data in a line plot. Point to the number with the most Xs, and say, *The most students have _____ letters in their* name. Have students repeat chorally. Point to the number with the fewest Xs, and say, *The fewest students have _____ letters in their name.* Have students repeat chorally. | **Developing Oral Language** Write a, e, i, o, and u on the board. Say, *The letters a, e, i, o, and u are vowels. My last name has _____ vowel(s).* Give each student a slip of paper. Direct them to write their last name and circle the vowels. Then have students count how many vowels and write the number. Collect the slips of paper, and use the data to make a line plot. Encourage students to analyze the data. Provide the following sentence frames: **Most students have _____ vowel(s). One student has _____ vowel(s). More students have _____ vowel(s) than _____ vowel(s).** | **Share What You Know** Create a line plot with increments of 0–10. Place Xs above numbers as follows: 4 has 3 Xs; 5 has 4 Xs; 6 has 8 Xs; 7 has 2 Xs; 8 has 3 Xs. Label the line plot, Windows. Explain that this line plot shows the number of windows on the fronts of houses on one city block. Ask questions to prompt students to analyze the data. Encourage students to answer in complete sentences. |

**Teacher Notes:**

NAME _____ DATE _____

# Lesson 8 Four-Square Vocabulary
## *Analyze Line Plots*

Write the definition for *line plot*. Write what the term means, draw a picture, and write your own sentence using the term.

| **Definition** | **My Own Words** |
|---|---|
| A graph that shows how often a certain number occurs in data. | See students' examples. |
| **My Picture** | **My Sentence** |
| See students' drawings. | I can use a line plot to show information about a survey. |

**line plot**

 **Teacher Directions:** Provide a description, explanation, or example of the new term using images or real objects. Have students use the Glossary to write the definition. Direct students to write a definition in their own words and draw a picture representing their math term. Have students write a sentence using the term. Then encourage students to read their sentence to a peer.

# Chapter 10 Time
## What's the Math in This Chapter?

### Mathematical Practice 2: Reason abstractly and quantitatively

Distribute manipulative clocks and present students with this scenario. *I need to be at the dentist by 4:00 pm. School is over at 3:00. It takes me half an hour to drive from school to the dentist. Will I be able to make it to the dentist on time?*

Have students manipulate their clocks to show 3:00. Then have students manipulate their clocks to show 4:00. Ask, *How much time do I have to get to the dentist?* Have students suggest various ways to figure out if there is enough time to drive to the dentist. Ask, *How much time is between 3:00 and 4:00?* **One hour or sixty minutes**

Ask, *What time do I need to leave to be at the dentist on time?* Allow students time to think and discuss. The goal of the discussion is for students to recognize and be able to show that you could leave anytime between 3:00 and 3:30 and still make it to the dentist on time. Point out to students that they reasoned abstractly and quantitatively as they talked/modeled about the meaning of the units of time in minutes and hours; therefore, they applied Mathematical Practice 2.

Display a chart with Mathematical Practice 2. Restate Mathematical Practice 2 and have students assist in rewriting it as an "I can" statement, for example: **I can reason abstractly and quantitatively to solve problems about time.** Post the new "I can" statement.

## Inquiry of the Essential Question:

### How do I use and tell time?

Inquiry Activity Target: **Students come to a conclusion that they can solve time related problems by thinking about the units of time.**

As an introduction to the chapter, present the Essential Question to students. The inquiry graphic organizer will offer opportunities for students to observe, make inferences, and apply prior knowledge of problem solving representing the Essential Question. As they investigate, encourage students to draw, write, and collaborate with peers to demonstrate their observations and thinking. Then have students present additional questions they may have to a peer to extend discussions.

Regroup students and restate Mathematical Practice 2 and the Essential Question. Pose questions to reflect on what has been learned to guide students in making connections between the Mathematical Practice and the Essential Question.

NAME _____ DATE _____

# Chapter 10 Time
## *Inquiry of the Essential Question:*

### How do I use and tell time?

Read the Essential Question. Describe your observations
(I see...), inferences (I think...), and prior knowledge (I know...) of
each math example. Write additional questions you have below.

I see ...

I think ...

I know ...

I see ...

I think ...

I know ...

half past 11

**Ride a Bike**

I see ...

I think ...

I know ...

A.M.
(P.M.)

Questions I have...

_____

_____

**Teacher Directions:** Read the Essential Question for students. Have students echo read.
Direct students to describe their observations, inferences, and prior knowledge of each
math example. Encourage students to write or draw additional questions they may have.
Then have students share their ideas/questions with a peer.

Grade 2 • Chapter 10 *Time* **83**

# Lesson 1 Time to the Hour

## *English Learner Instructional Strategy*

### Vocabulary Support: Multiple Meaning Words

Write *hour* and *minute* and their Spanish cognates, *hora* and *minuto* on the cognate chart. Have students refer to the glossary to write a definition and draw a labeled visual math example in their math journals. Model and say, *Point to your face. Say, face.* **face** *Show me your hands. Say, hands.* **hands** Stress the final /z/ sound. Display an analog clock and say, *A clock also has a face and hands.* Identify each term modeling with the clock. Have students repeat vocabulary chorally. Direct students to draw an analog clock in their math journals and label the math terms: *face* and *hands.*

### English Language Development Leveled Activities

| Emerging Level | Expanding Level | Bridging Level |
|---|---|---|
| **Word Recognition** | **Academic Vocabulary** | **Making Connections** |
| Display an analog clock. Point to the minute hand and say, *minute hand.* Repeat with the hour hand. Have students chorally repeat vocabulary. Then say, *two hands,* and have students repeat. Stress the singular and plural forms. Then ask word recognition questions, such as: *What is this? What are these?* Have students respond accordingly. **minute/hour hand; hands** Ask questions that require a response with a gesture for students not ready to verbally respond, such as: *Point to the hour/minute hand.* | Distribute manipulative analog clocks and write-on/wipe-off boards to students. On the board, write: 3 o'clock. Say, *Write how this time is shown on a* **digital** *clock.* Write 3:00 and have students compare their answers. Write: 11 o'clock. Say, *Show me this time on an* **analog** *clock. Where does the* **minute hand** *point?* **the twelve** *Where does the* **hour hand** *point?* **the eleven** Continue the activity, alternating between analog and digital clocks. | Distribute manipulative analog clocks to students. Set a demonstration digital clock to show 8:00. Then say, *Use the analog clock to show this time. Where does the hour hand point?* **The hour hand points to eight.** Ask, *Where does the minute hand point?* **The minute hand points to twelve.** Ask students to set their clocks to additional times and explain where each hand is pointing using complete sentences. |

### Multicultural Teacher Tip

Most ELs have had math education in their native countries and are familiar with basic math concepts. However, mathematical discourse in an unfamiliar language can be intimidating and confusing, and students may struggle with even seemingly simple steps leading to a solution. Manipulatives are a helpful option for EL students. By utilizing concrete objects to model familiar concepts or to learn new ones, students can work around language barriers that might make verbal or written explanations too difficult. Keep in mind that manipulatives are not always used in other cultures, and the student may need time and encouragement to become comfortable using them instead of solving on paper.

NAME _____  DATE _____

# Lesson 1 Note Taking
## *Time to the Hour*

Read the question. Write words you need help with. Use your lesson to write your Cornell notes.

| Building on the Essential Question | Notes: |
|---|---|
| **Building on the Essential Question**<br><br>How can I tell time to the hour? | <br><br>This is an __analog__  __clock__.<br><br>There are two __hands__ on an analog clock.<br><br>The __hour__ hand points to the hour. It is __shorter__.<br><br>The __minute__  __hand__ points to the minute. It is __longer__.<br><br><br><br>This is a __digital__  __clock__.<br><br>A digital clock shows __hours__ and __minutes__. |
| **Words I need help with:**<br>See students' words. | |

 **Teacher Directions:** Read the Building on the Essential Question and have students list words/phrases they need assistance with. Provide descriptions, explanations, or examples of the terms using images or real objects. Read each sentence frame and have students write the appropriate terms. Have students read their notes aloud.

# Lesson 2 Time to the Half Hour

## English Learner Instructional Strategy

### Language Structure Support: Communication Guide

Post and model using the following communication guide for students to utilize during this lesson.

This is a(n) _____ clock.
It is _____ o'clock.
The hour hand is pointing to _____.
The minute hand is pointing to _____.
One half hour is _____ minutes.
The time is _____ past _____.

### English Language Development Leveled Activities

| Emerging Level | Expanding Level | Bridging Level |
|---|---|---|
| **Building Word Knowledge** | **Building Oral Language** | **Act It Out** |
| Cut out and distribute both circles and half circles to each student. As you give them to students, describe each as "half" or "whole" correspondingly. Display a demonstration analog clock set to four o'clock. Move the minute hand around the clock once to reset the clock to five o'clock. Say, *whole hour.* Then move the minute hand and hour hand to set the clock to five-thirty and say, *half hour.* As you set the clock to show hours and half hours, have students display the whole circle or the half circle to demonstrate understanding. | Divide students into two groups. Provide one group with the sentence frame: **It is half past _____.** Provide the other group with the sentence frame: **It is _____ thirty.** Display a demonstration analog clock. Set the time to show a half hour, and have the first group use their assigned sentence frame to chorally tell what time is shown. Then repeat with the second group. Reset the clock several times and repeat. Have groups switch sentence frames to give students a chance to practice telling time to the half hour both ways. | Divide students into two groups. Distribute an analog clock to each student. Write and say, *1:00* and direct one group to show 1:00 on their clocks. Then direct the other group to show the next half hour on their clocks. **1:30** Then direct the first group to show the next half hour. **2:00** Continue on in this manner until students' understanding is firm. Repeat the activity having groups alternate showing the next hour or the next 2 hours. |

**Teacher Notes:**

NAME _____  DATE _____

# Lesson 2 Concept Web
## *Time to the Half Hour*

Circle the correct words.

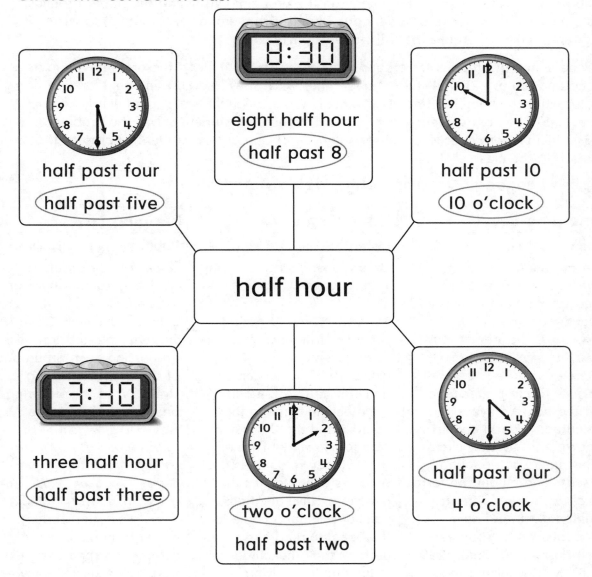

half past four

(half past five)

eight half hour

(half past 8)

half past 10

(10 o'clock)

half hour

three half hour

(half past three)

(two o'clock)

half past two

(half past four)

4 o'clock

**Teacher Directions:** Review the new terms. Model and prompt the following question and answer: **What time is it? It is three-thirty/It is half past three.** Have students circle each correct answer independently. Then have them compare with a partner: one student should ask **What time is it?** and the other student should answer.

Grade 2 • Chapter 10 *Time* **85**

# Lesson 3 Problem Solving Strategy: Find a Pattern

## English Learner Instructional Strategy

### Collaborative Support: Round the Table

Place students in multilingual groups of 4 or 5. Assign one problem from Exercises 1–6 to each group. Have students work jointly on the problem by passing the paper around the table for each member to provide input. Direct each member of the group to write with a different color to ensure that all students participate.

Provide a step-by-step list for groups to follow, such as: 1) Read the problem aloud as a group and discuss. 2) One student underlines what they know. 3) The next student circles what they need to find. 4) The next student highlights all signal words and phrases. 5) The next student writes a plan. 6) The next student solves the problem. 7) The last student checks for reasonableness. 8) Choose one student to present the solution to the class.

### English Language Development Leveled Activities

| Emerging Level | Expanding Level | Bridging Level |
|---|---|---|
| **Modeled Talk** | **Listen, Write, and Read** | **Public Speaking Norms** |
| On the board, write: *Understand, Plan, Solve,* and *Check.* Write then read aloud, *Each half hour, a bus arrives at the station. The first bus arrives at 8:00. When does the fourth bus arrive?* Circle *Understand* on the board. Have a volunteer underline what is known and another circle what needs to be found. Discuss and model the term, *each half hour,* with an analog clock and have students verbally say each half hour: **8:00, 8:30, 9:00, 9:30.** Model solving the problem with this pattern. | On the board, write then say, *It is half past 3 now. In two hours, it will be dinner time. What time will dinner be served?* Write each problem-solving step and an explanation of the step on an index card. Ask a pair of volunteers to come to the board. Hand the Understand card to one student. He or she will read it aloud as the second student follows the directions. Have additional pairs of volunteers come to the board to read and perform the other three steps in the problem-solving process. Encourage students to use the find a pattern strategy to solve the problem. | Draw a large problem-solving graphic organizer labeled: Understand, Plan, Solve, and Check. Divide students into four groups, and assign each group one of the problem-solving steps. Write the following on the board, then say, *Each TV show is half an hour long. Tim's favorite show is the fifth one on. It is now 5 o'clock. What time does Tim's favorite show start?* Have the students in each group direct you in completing their assigned step. Guide students toward the correct pattern needed to solve the problem. |

**Teacher Notes:**

NAME _____ DATE _____

# Lesson 3 Problem Solving
## *STRATEGY: Find a Pattern*

<u>Underline</u> what you know. (Circle) what you need to find. Find a pattern to solve.

**1.** In the morning, **Ms. White's** students **change activities every hour**.

**Reading** starts at **8:30**.

Her students **go through three more** activities.

reading

(When do each of the activities start?)

8:30, _____, _____, _____

The activities start at 8:30, <u>9:30</u>, <u>10:30</u> and <u>11:30</u>.

---

**2.** Students **start** work in **learning stations** at **10:30**.

**After** two hours, they **go** to **recess**.

recess

(What time do students go to recess?)

Students go to recess at <u>12:30</u>.

**Teacher Directions:** Provide a description, explanation, or example of the boldface terms and nouns using images or real objects. Read each sentence and have students echo read. Encourage students to find a pattern and then write their answers in each restated question. Have students read each answer sentence aloud.

# Lesson 4 Time to the Quarter Hour

## English Learner Instructional Strategy

### Vocabulary Support: Choral Responses

Write *quarter* on the board. Show circle fraction models representing one-quarter and one-half. Explain that one-fourth of a circle can also be called a quarter of a circle. Model and compare an analog clock at fifteen-minute intervals with the two circle fractional models. Discuss the terms: *quarter past, quarter till, and half past.*

On a digital clock, display a time to a fifteen-minute interval. Ask, *What time does the clock show?* Students will answer chorally with one of the following sentence frames: **The clock shows a quarter past _____. The clock shows a quarter till _____. The clock shows a half past _____. The clock shows _____ o'clock.**

### English Language Development Leveled Activities

| Emerging Level | Expanding Level | Bridging Level |
|---|---|---|
| **Word Knowledge** | **Number Game** | **Internalize Language** |
| Show students a circle divided into quarters. Count the quarters. Show an analog clock. Starting at 12, move the minute hand in fifteen-minute intervals. As you do, count the quarters. Ask, *How many quarters in one hour?* **4** Move the minute hand around the clock in 15-minute intervals again. Say, *One quarter is 15 minutes. Two quarters is 30 minutes. Three quarters is 45 minutes. How many is four quarters?* **60 minutes** Distribute manipulative analog clocks and have students repeat the activity with a partner. | Divide the students into four teams. Distribute an analog clock to each student. On a digital clock, display a time to a fifteen-minute interval. Say, *Show me this time as quickly as you can. The first team with all its members showing the correct time scores a point.* Repeat with various times at fifteen-minute intervals. The first team to score three points wins. | Invite a pair of students to the front. Use an analog clock to display a time at a fifteen-minute interval, for example 4:45. Ask the first student, *What time is it?* Have the student answer, saying for example, **four forty-five.** Then say to the second student, *Tell me the time in a different way.* The second student would say, **a quarter till five.** Invite another pair of students to the front. Set the clock to a new time at a fifteen-minute interval, and repeat the activity. Then have students work in groups of 3, rotating jobs as time setter and time tellers to repeat the activity. |

### Teacher Notes:

NAME _____ DATE _____

# Lesson 4 Four-Square Vocabulary
## *Time to the Quarter Hour*

Write the definition for *quarter hour*. Write what the term means, draw a picture, and write your own sentence using the term.

| | |
|---|---|
| **Definition**<br><br>A quarter hour is 15 minutes. Sometimes called quarter past or quarter till. | **My Own Words**<br><br>See students' examples. |
| **My Picture**<br><br>See students' drawings. | **My Sentence**<br><br>I ride the bus for a quarter hour after school. |

**quarter hour**

**Teacher Directions:** Provide a description, explanation, or example of the new term using images or real objects. Have students use the Glossary to write the definition. Direct students to write a definition in their own words and draw a picture representing their math term. Have students write a sentence using the term. Then encourage students to read their sentence to a peer.

Grade 2 • Chapter 10 *Time* **87**

# Lesson 5 Time to Five-Minute Intervals

## English Learner Instructional Strategy

### Graphic Support: Number Lines

Create 3 to 4 number lines showing increments of five up to the number 60. Purposely omit some numbers. Have students take turns finding the missing numbers as they count by fives. Have students create their own number lines with missing numbers and challenge a partner to find the missing five-minute intervals.

### English Language Development Leveled Activities

| Emerging Level | Expanding Level | Bridging Level |
|---|---|---|
| **Act It Out** | **Sentence Frames** | **Building Oral Language** |
| Display a demonstration analog clock. Discuss and model that it takes five minutes for the minute hand to move from one number to the next number. Distribute analog clocks and have students first count the minute tick marks between 12 and 1. **5 minutes** Point to the corresponding clock numbers and count by fives from 5 to 60 with students. Have students repeat using their manipulative clocks to model as they count chorally as a group. Encourage students to then repeat counting on their own as you listen in. | Display a demonstration analog clock. Set the hands to show times at five minute intervals. Have students explain how they know what time it is, based on where the hands are located. First model, then encourage students to use the following sentence frame: **I know it is ____ because the minute hand is on ____ and the hour hand is between ____ and ____.** | Pair students and distribute a manipulative analog clock to each student. The first student will set his or her clock to a time at a five minute interval without showing the clock to the other student. The first student will then describe the clock using the sentence frame: **The minute hand is on ____ and the hour hand is between ____ and ____.** The second student will follow the directions, set his or her clock to the same time, and then say aloud the time shown. Students compare clocks, switch roles, and repeat the activity. |

**Teacher Notes:**

NAME _____ DATE _____

# Lesson 5 Guided Writing
## *Time to Five-Minute Intervals*

**How do you tell time to five-minute intervals?**

Use the exercises below to help you build on answering the Essential Question. Write the correct word or phrase on the lines provided.

1.  **What key words do you see in the question?**
    tell, time, five-minute, intervals

2.  **What is the long hand on an analog clock?** __minute__ __hand__

3.  **Each mark on a clock face is one** __minute__.

4.  **How long does it take for the minute hand to move from 1 to 2?**
    __5__ minutes

5.  **I can** __skip__ __count__ **by 5s to tell the time.**

6.  **How do you tell time to five-minute intervals?**

    Sample answer: I look at what number the minute hand is on. Then I

    skip count by 5s until I get to the minute hand. If the minute hand is on

    the 4, I count 5, 10, 15, 20. So the time is 20 minutes past the hour.

**Teacher Directions:** Read the Building on the Essential Question. Provide descriptions, explanations, or examples of the terms using images or real objects. Read each sentence frame and have students use the lesson and Glossary to write the appropriate terms. Have students read the sentences aloud.

# Lesson 6 A.M. and P.M.

## English Learner Instructional Strategy

### Sensory Support: Pictures/Photographs

Review English terms from the lesson and exercises which may be unfamiliar to students, such as: *dream, bedtime, story, breakfast, wash the dog, swimming, soccer practice, sleeping/bed, dinner, lunch, lacrosse practice, library, party,* and *invitation* to provide additional language support. Display labeled pictures or photos to clarify word meaning and place in a prominent place in the classroom for students to reference during the lesson.

Pair emerging/expanding students with bilingual (same native language) bridging level students during the Problem Solving exercises. Allow use of native language to clarify meaning.

### English Language Development Leveled Activities

| Emerging Level | Expanding Level | Bridging Level |
|---|---|---|
| **Act It Out** | **Building Oral Language** | **Public Speaking Norms** |
| Use a demonstration analog clock and say, A.M. *starts at midnight. We sleep during the* A.M. *and then we wake up in the morning. The first part of the day until lunch is* A.M. Set the clock to show a time. Display a picture of a child waking up. Have students act out waking up. Repeat with other A.M. activities. Say, *The second part of the day is* P.M. *It starts at noon. Afternoon until dinner is* P.M. *We go to bed in the* P.M. Display pictures of families eating dinner and children going to bed. Then have students act out the activity. Repeat with other P.M. activities. | Create a two-column chart labeled A.M. and P.M. Divide the students into two groups and give a manipulative analog clock to each group. Have one group set its clock to a time showing a fifteen minute interval. Ask the second group, *What would you do at this time in the* A.M.? Have a volunteer answer with the sentence frame: **In the** A.M.**, I** _____. Record the answer. Repeat with P.M. Have groups switch roles and repeat the activity. Discuss if there are any activities that could be done in the A.M. and in the P.M. | Give each student six sticky notes. Tell them to write or draw a picture of three activities that occur during the A.M. and three activities that occur during the P.M. Have students also write a corresponding time for each activity, designating either A.M. or P.M. Make a two-column chart labeled A.M. and P.M. Have students present their activities to the class, and then place their sticky notes on the chart in the appropriate column. After all students have presented their activities, sequence the activities in order according to specific hours of the day. |

**Teacher Notes:**

NAME _____ DATE _____

# Lesson 6 Vocabulary Definition Map
## *A.M. and P.M.*

Use the definition map to write what the math word means and tell what the word is like. Write or draw a math example. Share your examples with a classmate.

**My Math Words:**

A.M. and P.M.

**What It Is Like:**

12 A.M. is __midnight__ .

**What It Means:**

A.M.: Hours from midnight until noon.

P.M.: Hours from noon until midnight.

12 P.M. is __noon__ .

A.M. is morning.
P.M. is __afternoon__ .

**My Math Example:**

See students' examples.

**Teacher Directions:** Provide a description, explanation, or example of the new term using images or real objects. Have students use the lesson or Glossary to define the math term. Teach the meaning of the compound word *afternoon*. If students remember that noon is 12 P.M., then anything after that time is "after noon" or *afternoon*. Direct students to list characteristics and draw a picture representing their math terms. Then encourage students to describe their picture to a peer.

Grade 2 • Chapter 10 *Time*  89

# Chapter 11 Customary and Metric Lengths

## What's the Math in This Chapter?

### Mathematical Practice 6: Attend to precision

Distribute write-on/wipe-off boards and show students a piece of string or yarn that is 12 inches long. Say, *This is 1 long.* Ask students to draw a line that is the same length on their write-on/wipe-off board. After students have had time to draw their lines, go around the group and compare your string to their drawings.

Say, *I see many of you have lines that are different lengths. I should have been more precise and said that I wanted you to draw a line that was 1 **foot** long?* Have students redraw their lines using a ruler to be precise with their measurement. Discuss that it is important to determine a unit of measure.

Discuss with students how there are many units of measurement. Model and describe various measurement tools. Say, *Think about items that you have measured in the past. Turn and talk with a friend about the items you measured and how you measured them. Be precise.* The goal of this discussion should be for students to identify a measurable object and be precise and give a unit of measurement so others can understand what they mean.

Display a chart with Mathematical Practice 6. Restate Mathematical Practice 6 and have students assist in rewriting it as an "I can" statement, for example: **I can be precise when I talk about measurements.** Post the new "I can" statement.

## *Inquiry of the Essential Question:*

### How can I measure objects?

Inquiry Activity Target: **Students come to a conclusion that measurement can be represented in different ways.**

As an introduction to the chapter, present the Essential Question to students. The inquiry graphic organizer will offer opportunities for students to observe, make inferences, and apply prior knowledge of measurement representing the Essential Question. As they investigate, encourage students to draw, write, and collaborate with peers to demonstrate their observations and thinking. Then have students present additional questions they may have to a peer to extend discussions.

Regroup students and restate Mathematical Practice 6 and the Essential Question. Pose questions to reflect on what has been learned to guide students in making connections between the Mathematical Practice and the Essential Question.

NAME _____ DATE _____

# Chapter II Customary and Metric Lengths
*Inquiry of the Essential Question:*

## How can I measure objects?

Read the Essential Question. Describe your observations (I see...), inferences (I think...), and prior knowledge (I know...) of each math example. Write additional questions you have below.

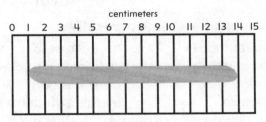

The craft stick is 13 centimeters long.

I see ...

I think ...

I know ...

Find the object. Measure it.
Tool: centimeter ruler

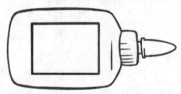

I see ...

I think ...

I know ...

Measure: **about** _____ centimeters

Questions I have...

_____

_____

**Teacher Directions:** Read the Essential Question for students. Have students echo read. Direct students to describe their observations, inferences, and prior knowledge of each math example. Encourage students to write or draw additional questions they may have. Then have students share their ideas/questions with a peer.

# Lesson 1 Inches

## English Learner Instructional Strategy

### Collaborative Support: Native Language Peers

Add *estimate, inch, length,* and *measure* to the Math Word Wall. For Exercises 1–7, have students partner with a native English speaker. Preteach language for students to use, such as: *What is this called? Is this correct? What do you think? Do you understand this? What is your estimate? I think the length is _____.* Emphasize to the English-speaking students that they should allow a little extra time for their peers to construct sentences.

### English Language Development Leveled Activities

| Emerging Level | Expanding Level | Bridging Level |
|---|---|---|
| **Choral Responses** | **Building Oral Language** | **Sentence Frames** |
| Write inch on the board. Display a color tile and a ruler. Say, *I am going to* **measure** *this tile.* Model using the ruler to measure the tile. Point to inch on the board as you say, *This tile is one* **inch** *long. How many inches is the tile?* Have students answer chorally, **one**. Say, *I will measure with tiles.* Measure other objects using color tiles. Count the tiles, saying, *one inch, two inches, three inches,* and so on. Then ask, *How many inches?* Have students answer chorally. Distribute tiles to students. Have them measure objects such as their desktops or shoes. | Provide each student an object to measure with a ruler. Say, ***Estimate*** *the object's length.* Have students record their estimates. Distribute rulers to students. Have students measure the object and record their measurements. Direct students to ask their partners: **How long is the _____?** Have the students answer, **The _____ is _____ inches long.** Direct students to trade objects with another pair and repeat the activity. | Provide each student an object to measure. Say, ***Estimate*** *the object's length.* Have students record their estimates and share them with the class using the sentence frame: **I estimated that the _____ would be _____ inches long.** Distribute rulers to students. Say, *Measure the object.* Have students measure, and then share the measurement using the sentence frame: **I measured the _____, and it is _____ inches long.** |

**Teacher Notes:**

NAME _____  DATE _____

## Lesson I Vocabulary Sentence Frames
### *Inches*

The math words in the box are for the sentences below. Write the words that fit in each sentence on the blank lines.

| estimate | length | measure |
|----------|--------|---------|

**I.** ___Length___ is how long something is.

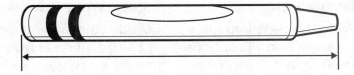

**2.** To ___estimate___ is to find a number that is **close** to an exact amount.

**3.** You can use a ruler to ___measure___ an object.

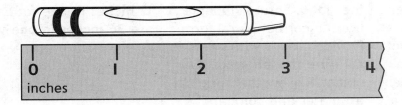

 **Teacher Directions:** Provide a description, explanation, or example of the each term using images or real objects. Read each sentence frame and have students echo read. Direct students to write the correct terms in each blank. Then encourage students to read each sentence to a peer.

Grade 2 • **Chapter II** *Customary and Metric Lengths*  **91**

# Lesson 2 Feet and Yards

## English Learner Instructional Strategy

### Vocabulary Support: Draw Visual Examples

Add the terms *foot*, and *yard* (*yarda*, Spanish cognate) to the Math Word Wall. Organize students into groups of 3–5 and assign a term to each group. Have students look up the term in the Glossary. Provide students with chart paper to write a definition, and draw a visual math example with the term labeled. Then, invite a volunteer from each group to share their definitions using the sentence frame: **Our math word is _____. It means _____.** Post the charts in the classroom, and have students record the terms and definitions in their math journals for future reference.

### English Language Development Leveled Activities

| Emerging Level | Expanding Level | Bridging Level |
|---|---|---|
| **Making Connections** | **Number Game** | **Share What You Know** |
| Provide each student with 12 color tiles. Have students get into groups of three. Write foot and yard on the board. Say, *Twelve inches is equal to one foot. Use tiles to make one foot.* Distribute rulers to students. Have them measure their tiles with their rulers. Discuss that each tile is one inch in length. Have students connect their 12 tiles. Display a yardstick. Explain and model one yard. Have each group of three students connect and count their three sets of tiles. **36** Ask groups to measure their connected tiles. Discuss that a yard is equal to 3 feet or 36 inches. | Divide the class into four groups named 1 Foot, 2 Feet, 3 Feet, and 4 Feet. Direct each group to find an object in the classroom that is about the same number of feet long as the name of their group, and to stand by the object, or bring the object back to their seats. Have volunteers describe the objects' lengths using the sentence frame: **The _____ is about _____ foot/feet long.** Ensure students use the correct plural form. Provide groups with rulers to measure the length of their object. Review their original estimates and compare them with their measurements. | Divide students into groups of three. Randomly assign/name each group as: Inch, Foot, or Yard. Tell each group to write three sentences to describe their unit of measurement. Allow students several minutes to write their descriptions. Have volunteers from each group share their descriptions. For example, **One inch is about as long as a paper clip. There are 12 inches in one foot. There are 36 inches in one yard.** |

### Teacher Notes:

NAME _____  DATE _____

# Lesson 2 Multiple Meaning Word
## *Feet and Yards*

Spell then say the math word. Draw a picture that shows the math word meaning and non-math word meaning in the boxes.

| foot | |
|---|---|
| **Math Meaning** | **Non-Math Meaning** |
| Students' examples should represent a customary unit for measuring length. | See students' examples. |

Use the sentence frame below to help you describe your pictures.

This picture for the word ___foot___ shows _____.

**Teacher Directions:** Provide math and non-math descriptions, explanations, or examples of the new term using images or real objects. Have students spell then say the term. Then direct students to draw pictures showing a math and non-math meaning of the math term. Encourage students to describe their pictures to a peer using the sentence frame.

# Lesson 3 Select and Use Customary Tools

## English Learner Instructional Strategy

### Sensory Support: Realia

Before the lesson, write: *ruler, yardstick*, and *measuring tape* on the board. Introduce each tool and allow students to look at and hold the actual tools. Have students write the terms and draw a visual example in their math journals. Next, have students use the Glossary to write definitions and sentences using each term in context. Provide sentence examples if needed. Accommodate non-Spanish speaking ELs with a translation tool.

Post and model the following sentence frames to assist students during the lesson: **I would measure the _____ [item] using a _____ [tool].**

### English Language Development Leveled Activities

| Emerging Level | Expanding Level | Bridging Level |
|---|---|---|
| **Activate Prior Knowledge** | **Word Lists** | **Report Back** |
| Display a ruler, yardstick, and measuring tape. Then show a photograph of a bus or other large object. Say, *I want to **measure** a bus. Should I use a **ruler**?* Prompt students to answer, **no** chorally or with a gesture. Ask, *Should I use a **yardstick**?* **no** Say, *The bus is longer than a ruler. The bus is longer than a yardstick. I should use a **measuring tape**.* Use other objects and photographs as you repeat the activity, guiding students to recognize the correct tool for measuring. | Divide students into three groups. Randomly assign each group to be a Ruler, a Yardstick, or a Measuring Tape. Say, *List four or more things you would measure with your assigned tool.* Allow groups several minutes to generate their lists. Have students from each group use complete sentences to share the objects they listed. Provide the following sentence frame: **I would use a _____ to measure a _____.** | Divide students into three groups. Provide a ruler, yardstick, and measuring tape to each group. Say, *Use each tool to measure something in the classroom. You will need to find one object that is the right length to be measured with each tool.* Allow students several minutes to choose their objects and measure them. Have each group present a short report. They should present the object they measured, describe the tool they used, and say how long the object is. As a group, discuss whether the tool was the right choice for the object measured. |

### Teacher Notes:

NAME _____   DATE _____

# Lesson 3 Word Identification
## *Select and Use Customary Tools*

Match.

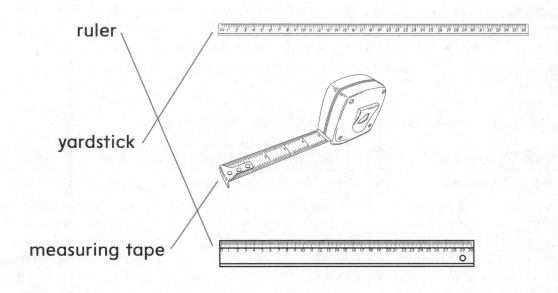

ruler

yardstick

measuring tape

---

Write the correct term from above for each sentence on the blank lines.

**I.**  I would use a __measuring__ __tape__ to find the length of a car.

**2.**  I would use a __ruler__ to find the length of my hand.

**3.**  I would use a __yardstick__ to find the length of my desk.

**Teacher Directions:** Review the terms using images or real objects. Have students say each term then draw a line to match the term to its picture. Direct students to say each sentence and then write the corresponding meanings in the sentences. Encourage students to read the sentences to a peer.

**Grade 2 · Chapter II** *Customary and Metric Lengths*  **93**

# Lesson 4 Compare Customary Lengths
## English Learner Instructional Strategy

### Vocabulary Support: Modeled Talk

Write long and longer on the board. Display two objects of different lengths. Model and say for example, *This pen is long. This book is longer.* Point to each term in turn as you say *long* and *longer* and stress the suffix -er. Have students repeat chorally. Repeat with the terms *short* and *shorter*.

Ask students to take an object out of their desk. Then direct students to find an object that is longer than the first one and place both objects on their desk. Have students turn and talk with a neighbor using these sentence frames: **This _____ [item name] is longer. This _____ [item name] is shorter.**

### English Language Development Leveled Activities

| Emerging Level | Expanding Level | Bridging Level |
|---|---|---|
| **Word Recognition** | **Listen, Write, and Read** | **Phonemic Awareness** |
| Randomly distribute two to twelve connecting cubes to each student. Say, *Join your connecting cubes.* Invite a pair of students to display their cube trains, one train horizontally next to the other. Use comparisons to ask questions about the cube trains. For example, point to the shorter cube train, and ask, *Is this cube train longer than that train?* Have students answer **no** chorally or with a gesture. Repeat with different pairs of students. | Write *longer* and *shorter* on the board. Display two items of different lengths. Say, *Use a word from the board to write a description.* Have students write a complete sentence describing the lengths of the items. For example, **The book is longer than the pencil. The pencil is shorter than the book.** Have students read their sentences to the group. Repeat with new items. To extend language, measure the items and have students incorporate the measurements into their sentence. For example, **The crayon is _____ inches shorter than the pen** | Divide students into pairs. Distribute a ruler and two objects to each pair. Have them measure each object and record the lengths to the nearest inch. Say, *Write sentences comparing the lengths of your objects. Use the words **longer** and **shorter** correctly in your sentences.* Give students several minutes to write their sentences. Then have pairs share their comparisons. For example, **The _____ is _____ inches shorter than the _____.** Be sure students are correctly saying the /ch/ sound in *inches.* |

### Teacher Notes:

NAME _____   DATE _____

# Lesson 4 Concept Web
## *Compare Customary Lengths*

Circle the correct word for the circled object.

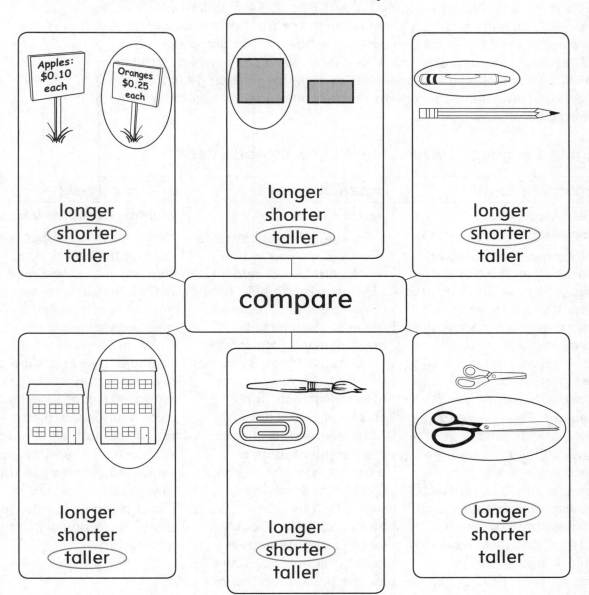

**Apples: $0.10 each**   **Oranges $0.25 each**

longer
(shorter)
taller

longer
shorter
(taller)

longer
(shorter)
taller

compare

longer
shorter
(taller)

longer
(shorter)
taller

(longer)
shorter
taller

**Teacher Directions:** Review the terms *compare, longer, shorter,* and *taller*. Identify and explain each object discussing the comparative images and using real objects. Have students individually complete the web circling the best description for each object. Then have partners compare answers using the sentence frame: **The _____ is [longer/ shorter/taller] than the _____.**

# Lesson 5 Relate Inches, Feet, and Yards

## English Learner Instructional Strategy

### Sensory Support: Act It Out

On each of nine index cards, write one of the following: *inches, feet, yards, ruler, yardstick, measuring tape, textbook, door, hallway*. Distribute cards to 9 students and have them come to the front of the classroom. Ask the class to read the cards and determine how best to group the terms. Direct students to turn and talk with a peer. Once pairs have determined which object, tool, and measurement go together in each group, discuss as a class. Have standing students rearrange themselves into the 3 determined groups.

### English Language Development Leveled Activities

| Emerging Level | Expanding Level | Bridging Level |
|---|---|---|
| **Build Background Knowledge**<br><br>Write *measure, ruler,* and *yardstick* on the board. Display a ruler and yardstick. Say, *We use these tools to measure objects.* Show the ruler, and ask, *Is this a ruler?* Have students answer **yes** chorally or with a gesture. Show the yardstick, and say, *This is a **yardstick**.* Then ask, *Is this a ruler?* Have students answer, **no**. Encourage students to repeat, **This is a yardstick.** Display objects that would be measured by either a ruler or yardstick. For each object, ask, *What should I use to measure this ____?* Have students answer chorally or by pointing to the correct tool. | **Partner Work**<br><br>Have students work in pairs. Provide each pair with an index card. Direct students to write a key to help them remember the relationship between the units of measurement. For example, 12 inches = 1 foot, 36 inches = 1 yard, 3 feet = 1 yard. Measure objects in the classroom. Then have volunteers restate the measurement using a different unit of measure. Provide the following sentence frame: **The ____ is also ____ inches/feet/yards long.** Ensure students are saying the final /z/ sound for the plural form *inches* and *yards*. | **Multiple Word Meanings**<br><br>Write *inch, ruler, foot,* and *yard* on the board. Have students brainstorm multiple definitions for these words and demonstrate the meanings by using the words in sentences. For example: **I wear a shoe on my foot. I play soccer in my yard. A king is called a ruler. Inch forward to reach the salt.** Then have students use the words in sentences that demonstrate their math definitions. Direct students to refer to their My Vocabulary Cards if necessary. |

## Teacher Notes:

NAME _____ DATE _____

# Lesson 5 Guided Writing

*Relate Inches, Feet, and Yards*

**How do you relate inches, feet, and yards?**

Use the exercises below to help you build on answering the Essential Question. Write the correct word or phrase on the lines provided.

1. What key words do you see in the question?
   relate, inches, feet, yards
   _____

2. How many inches are in I foot? __12__

3. How many feet are in I yard? __3__

4. I can use different __units__ to measure the length of the same __object__.

5. I can use a __ruler__ to measure inches.

6. I can use a __yardstick__ to measure feet and yards.

7. How do you relate inches, feet, and yards?

   Sample answer: I can measure the same object in inches, feet
   _____

   and yards.
   _____

 **Teacher Directions:** Read the Building on the Essential Question. Provide descriptions, explanations, or examples of the terms using images or real objects. Read each sentence frame and have students use the lesson and Glossary to write the appropriate terms. Have students read the sentences aloud.

**Grade 2 • Chapter II** *Customary and Metric Lengths* **95**

# Lesson 6 Problem Solving Strategy: Use Logical Reasoning

## English Learner Instructional Strategy

### Collaborative Support: Echo Reading

Display photos of a garden, a football field (100 yards) and a high diving board at a pool to visually assist students with the Problem Solving and the Practice the Strategy examples. Read aloud each problem and have the class echo read.

Pair ELs with a native English-speaking student for Exercises 1–6. Direct the native English-speaking student to read the problems aloud, and have the EL echo read. Encourage the EL to ask their partner to explain terms they are unfamiliar with by asking, **What does this mean?**

### English Language Development Leveled Activities

| Emerging Level | Expanding Level | Bridging Level |
|---|---|---|
| **Activate Prior Knowledge** | **Sentence Frames** | **Logical Reasoning** |
| On the board, write: *Understand*, *Plan*, *Solve*, and *Check*. Then write and read aloud this word problem: *My bike is 3 feet long. My cousin's bike is 2 yards long. Whose bike is longer?* Underline the first two sentences. Say, *These are the things we know. Do we know which bike is longer?* Allow students time to respond, **no**. Circle the third sentence, and say, *I will circle what we need to find. This helps me understand the problem.* Point to *Plan*, and say, *Now we need a plan.* Continue in this manner as you model solving the problem. | Draw four large boxes on the board, and label them: Understand, Plan, Solve, and Check. In the Understand box, write, *An ice sculpture is 2 feet tall. It melts after an hour and is shorter. Is it now 21 inches or 26 inches tall?* Provide sentence frames to help students guide you in completing the problem-solving graphic organizer. For example: **I know _____, so I will underline _____. I need to find out _____, so I will circle _____. I will _____ to solve the problem. I will _____ to check the answer.** | Draw a large problem-solving graphic organizer on the board with four large boxes labeled: Understand, Plan, Solve, and Check. Divide students into four groups, and give each group four index cards. On the board, write this word problem: *A man buys a tree at a garden center. He puts the tree in his car to drive it home. Would the tree be 4 inches tall, 4 feet tall, or 4 yards tall?* Have a student read the problem aloud. Direct groups to solve the problem and record each step on their four index cards. Then have each group describe one of the steps as you complete the graphic organizer on the board. |

**Teacher Notes:**

NAME _____ DATE _____

# Lesson 6 Problem Solving
## STRATEGY: Use Logical Reasoning

<u>Underline</u> what you know. (Circle) what you need
to find. Use logical reasoning to solve.

1. **Sam planted** a tomato plant that is
<u>**I foot tall**</u>.

The plant **grows** a little each week.

After **4 weeks**, would the plant be
10 inches **or** 14 inches tall?

tomato plant

The plant would be __14__ inches tall.

---

2. **Jane** made a paper chain that is
**I yard long**.

**Brad** made a paper chain that is
**2 feet long**.

Who made the longer paper chain?

paper chain

____Jane____ made the longer paper chain.

 **Teacher Directions:** Provide a description, explanation, or example of the boldface terms
and nouns using images or real objects. Read each sentence and have students echo read.
Encourage students to use logical reasoning and then write their answers in each restated
question. Have students read each answer sentence aloud.

# Lesson 7 Centimeters and Meters

## English Learner Instructional Strategy

### Vocabulary Support: Modeled Talk

Add the words *centimeter* and *meter* and their Spanish cognates *centímetro* and *metro* to the cognate chart. Some ELs may be very familiar with these words.

On the board, write: 100 centimeters = 1 meter. Show a centimeter ruler. Point at the marks as you say, *These are centimeters.* Distribute centimeter rulers to each student and have them identify the centimeters. Show a meterstick, and say, *This is one meter.* Then have students count with you by 10s to one hundred as you point to each increment of 10 on the meterstick. Say, *One meter is 100 centimeters long.* Emphasize *meter* and *centimeter* as you say them. Have students write the new vocabulary terms in their math journals with definitions as well as the equivalent: 100 centimeters = 1 meter.

### English Language Development Leveled Activities

| Emerging Level | Expanding Level | Bridging Level |
|---|---|---|
| **Build Word Knowledge** | **Building Oral Language** | **Explore Language Structures** |
| Use a centimeter ruler and a meterstick to measure objects in the classroom. Say the number, but ask students to identify the unit of measurement. For example, measure a book, then say, 27. *Is it 27 meters or 27 centimeters in length?* Give students a chance to respond chorally or by gesturing to the centimeter ruler or the meterstick. Then say, *The book is 27 centimeters in length.* Repeat with additional objects as time allows, encouraging students to respond with the correct terms as they get more practice. | Have students work in pairs. Distribute an inch ruler and a centimeter ruler to each pair. Then give each pair an object to measure. Ask each pair, *How many inches is the _____?* Have one student use the inch ruler to measure and answer in a complete sentence. Then ask, *How many centimeters is the _____?* Have the other student use the centimeter ruler to measure and answer in a complete sentence. Have pairs switch rulers, rotate objects among the groups and repeat the activity. | On the board, write: cent, centi, centipede, century, and centimeter. Say, *Cent- or centi- means one hundred. There are 100 cents in one dollar. How many feet does a centipede have?* **100** *How many years in a century?* **100** *How many centimeters are there in a meter?* **100** Have students underline *cent-* or *centi-* in each corresponding word on the board. Provide pairs of students a meterstick. Direct pairs to find something in the classroom that measures about 100 centimeters. List the found objects on a chart, then discuss their findings. |

### Multicultural Teacher Tip

As the metric system is the standard throughout most parts of the world, ELs will most likely be more familiar with units of metric measurement than they will be with standard units. Students who have worked only with the metric system in the past will be more familiar with partial amounts written as decimals, not fractions.

NAME _____ DATE _____

# Lesson 7 Note Taking
## *Centimeters and Meters*

Read the question. Write words you need help with. Use your lesson to write your Cornell notes. Write or draw math examples and explain your thinking to a classmate.

| **Building on the Essential Question** | **Notes:** |
|---|---|
| How can I measure in centimeters and meters? | I can <u>measure</u> to find the length of an object.<br><br>Use <u>centimeters</u> to measure shorter objects.<br><br>Use a centimeter <u>ruler</u> to measure in centimeters.<br><br>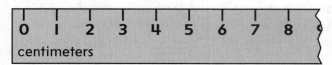<br><br>Use <u>meters</u> to measure longer objects.<br><br>Use a <u>meterstick</u> ruler to measure in meters.<br><br>There are <u>100</u> centimeters in a meter. |
| **Words I need help with:**<br>See students' words. | |
| **My Math Examples:**<br>See students' examples. | |

 **Teacher Directions:** Read the Building on the Essential Question and have students list words/phrases they need assistance with. Provide descriptions, explanations, or examples of the terms using images or real objects. Read each sentence frame and have students write the appropriate terms. Have students read their notes aloud. Direct students to draw a picture representing the question. Then encourage students to describe their picture to a peer.

# Lesson 8 Select and Use Metric Tools

## English Learner Instructional Strategy

### Sensory Support: Photographs/Realia

To introduce the lesson, use a meterstick to lead students in counting the 100 centimeters that make its length. Then review the photographs pictured in the exercises as you point to the corresponding objects in the classroom to help students learn the English terms: *door, easel, eraser, paper clip, bookshelf, rug, notebook,* and *paintbrush.* Write the terms on sentence strips and label each classroom object.

Pair emerging/expanding students with bilingual (same native language) bridging level students during the Problem Solving exercises. Allow use of native language to clarify meaning of unfamiliar terms.

### English Language Development Leveled Activities

| Emerging Level | Expanding Level | Bridging Level |
|---|---|---|
| **Graphic Support** Draw a T-chart and label it with: Small Objects and Big Objects. Ask students to give you 4 examples for each category that can be found in the classroom. Write the examples on the chart. Discuss whether you would use a centimeter ruler or meterstick to measure the items. Have students work in pairs. Provide each pair with a centimeter ruler and meterstick. Direct students to measure the items listed on the chart and record their measurements. Use the T-chart to record students' measurements and discuss as a group. | **Exploring Language Structure** Distribute small objects and pictures of large objects to students. Show one of the objects or pictures, and model using an if/then conditional phrase. For example, show a picture of a bus, and say, *If I want to measure a bus, then I should use a meterstick.* Have each student show his or her object and describe the tool needed to measure it. Provide the following sentence frame: **If I want to measure a ____, then I should use a ____.** | **Group Activity** Divide the students into two groups. Provide one group a meterstick and the other group a centimeter ruler. Say, *Find two objects in the classroom that you would measure with the tool you were given. Then measure them.* Allow groups several minutes to find the objects, measure them, and record the lengths in meter(s)/ centimeters. Then have each group share the items they measured and tell how long each object was in their assigned unit. For example, **We measured the table. It is about two meters long.** |

**Teacher Notes:**

NAME _____ DATE _____

# Lesson 8 Concept Web
## *Select and Use Metric Tools*

Circle the tool you should use to measure the object.

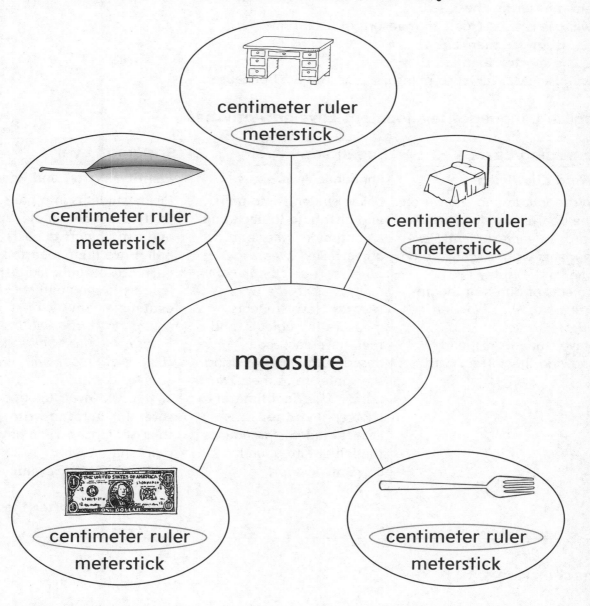

centimeter ruler
(meterstick)

(centimeter ruler)
meterstick

centimeter ruler
(meterstick)

measure

(centimeter ruler)
meterstick

(centimeter ruler)
meterstick

**Teacher Directions:** Review the terms *measure*, *centimeter ruler* and *meterstick*. Identify and explain each object using comparative images and real objects. Have students individually complete the web circling the best measurement tool for each object. Then have partners compare answers using the sentence frame: **I would measure a** _____ **with a** _____.

98    Grade 2 • **Chapter II** *Customary and Metric Lengths*

# Lesson 9 Compare Metric Lengths
## English Learner Instructional Strategy

### Vocabulary Support: Communication Guide

Post and model using the following communication guide for students to utilize during the measurement activities.

I am measuring the _____.
I will use a _____ [tool] to measure in _____ [unit].
_____ is longer than _____.
_____ is shorter than _____.
I will use _____ [unit] to measure _____ [object] because...

### English Language Development Leveled Activities

| Emerging Level | Expanding Level | Bridging Level |
|---|---|---|
| **Word Knowledge**<br><br>Write *longer* and *shorter* on the board. Provide two index cards to each student. Have students write each word on one card. Display two objects of different lengths. Ask, *Is the _____ longer or shorter than the _____?* Have students respond by showing the correct card. | **Phonemic Awareness**<br><br>Divide students into pairs, and distribute to each pair a centimeter ruler and two objects. Say, *Measure both objects. Then **compare** their lengths using the word **shorter**.* Have students measure the objects, and then have each pair describe the lengths using the sentence frames: **The _____ is _____ centimeters shorter than the _____.** Ensure students pronounce the initial /sh/ sound correctly in *shorter*. | **Identify, Write, and Read**<br><br>Divide students into pairs. Distribute a centimeter ruler and two objects to each pair. Have them measure each object and record the lengths to the nearest centimeter. Say, *Write sentences comparing the lengths of your objects. Use the words **longer** and **shorter** correctly in your sentences.* Give students several minutes to write their sentences. Then have pairs share their comparisons. For example, **The _____ is _____ centimeters shorter than the _____.** |

### Teacher Notes:

NAME _____ DATE _____

# Lesson 9 Four-Square Vocabulary
## *Compare Metric Lengths*

Write the definition for *compare*. Write what the word means, draw a picture, and write your own sentence using the word.

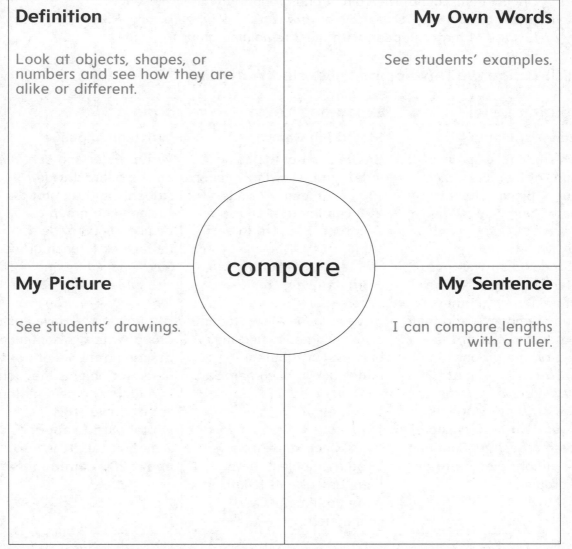

**Definition**

Look at objects, shapes, or numbers and see how they are alike or different.

**My Own Words**

See students' examples.

compare

**My Picture**

See students' drawings.

**My Sentence**

I can compare lengths with a ruler.

**Teacher Directions:** Provide a description, explanation, or example of the new term using images or real objects. Have students use the Glossary to write the definition. Direct students to write a definition in their own words and draw a picture representing their math term. Have students write a sentence using the term. Then encourage students to read their sentence to a peer.

Grade 2 • **Chapter II** *Customary and Metric Lengths* **99**

# Lesson 10 Relate Centimeters and Meters

## English Learner Instructional Strategy

### Sensory Support: Hands-On Activity

On the board, write: 100 centimeters = 1 meter. Divide the class into groups of four. Cut pieces of yarn or string 1 meter long and distribute one to each group. Say, *Each piece of yarn is 100 centimeters long, or 1 meter long. Find something in the room that is about 1 meter long. Use your piece of yarn to measure.* Discuss items found. Then direct one group to lay their piece of yarn out on the ground. Say, *One piece of yarn is 100 centimeters long. How many meters is it?* **1 meter** Repeat, with the 3 remaining groups.

### English Language Development Leveled Activities

| Emerging Level | Expanding Level | Bridging Level |
|---|---|---|
| **Listen and Identify** | **Measuring Game** | **Written Language** |
| Have students work in groups of four. Give each group 4 pieces of yarn 1 meter long. Say, *One piece of yarn is 100 centimeters long. How many meters is it?* Students answer, **one meter** chorally or with a gesture. Direct another group member to add their string to the end of the string on the ground. Say, *Two pieces of yarn are 200 centimeters long. How many meters is that?* Students answer, **two meters** chorally or with a gesture. Continue until all four pieces are on the ground. | Create a set of matching index cards showing 1 meter, 100 centimeters, 2 meters, 200 centimeters, and so on. Distribute one card to each student. Say, *Find the person whose card shows the same length as yours, but in a different unit of measurement.* Allow students a few minutes to find their partners. Distribute a meterstick to each pair. Say, *Find an object in the room that is about the same as the length written on your cards.* Once students have located an object, have them describe its length using both meters and centimeters. | Write *meter* and *centimeter* on the board. Divide students into six groups. Assign each group a number 1–6. Provide a meterstick to each group. Say, *Find something that is the same number of meters long as the number you were assigned.* Have each group write two sentences describing the length of the object or objects they found in meters and centimeters. For example, **This whiteboard is about 2 meters tall. It is also about 200 centimeters tall.** |

## Teacher Notes:

NAME _____ DATE _____

# Lesson 10 Note Taking
## *Relate Centimeters and Meters*

Read the question. Write words you need help with.
Use your lesson to write your Cornell notes. Write or
draw math examples to explain your thinking then
share with a classmate.

| **Building on the Essential Question** | **Notes:** |
|---|---|
| How can I relate centimeters and meters? | I can use different units of __length__ to measure the same object. |
| | I can measure the object in __meters__ with a meterstick. |
| | Then I can measure an object again in __centimeters__ with a centimeter ruler. |
| **Words I need help with:** | Measuring in centimeters might give a more __exact__ measurement. |
| See students' words. | |

**My Math Examples:**

See students' examples.

 **Teacher Directions:** Read the Building on the Essential Question and have students list words/phrases they need assistance with. Provide descriptions, explanations, or examples of the terms using images or real objects. Read each sentence frame and have students write the appropriate terms. Have students read their notes aloud. Direct students to draw a picture representing the question and describe their picture to a peer.

**100** Grade 2 • **Chapter 11** *Customary and Metric Lengths*

# Lesson 11 Measure on a Number Line

## English Learner Instructional Strategy

### Collaborative Support: Mentor/Aide

Write *number line* and its Spanish cognate *línea numérica* on the cognate chart.

Before the lesson, have a native language mentor/aide work with students to introduce measuring with a number line. Have the mentor model and talk students through the process shown on the Explore and Explain page.

To provide additional language support, review the following terms using photos/drawings to clarify meaning for the exercises: *long/longer, short/shorter, tall/taller, brother, trail, woods.*

### English Language Development Leveled Activities

| Emerging Level | Expanding Level | Bridging Level |
|---|---|---|
| **Number Recognition** | **Report Back** | **Pairs Share** |
| Display a number line from 0 to 15. Say, *This is a number line.* Place an object on the number line, but do not line it up at zero. For example, place a pencil on the number line so it starts at 3 and ends at 9. Count to measure the pencil, but then ask, *Is the pencil nine units long?* Allow students time to answer **no** either chorally or with a gesture. Then say, *You are right. The pencil is not nine units long. The pencil starts at 3 and ends at 9. It is six units long.* Repeat with other objects. Allow students to model measuring using the number line. | Have students work in pairs. Distribute blank number lines with one-inch increments to each pair. Direct students to label the increments 12, 13, 14, 15, and so on. Then give each pair an object to measure. Say, *Put your object on the number line. Use the numbers to measure.* Have students measure. Then have a student from each pair say, **We used the number line to measure. The ____ is ____ inches long.** Ensure students are pronouncing the final /d/ sound in *used.* | Have students work in pairs. Distribute a blank number line showing one-inch increments to each pair. Say, *Starting with any number other than zero, write numbers in order on your number line.* Have pairs exchange number lines. Then give each pair an object to measure. After students have measured, have them describe the number line they were given and the length of the object. For example, **Our number line shows 36 to 46. The stapler is five inches long.** Rotate objects around the groups and repeat activity. |

## Teacher Notes:

NAME _____ DATE _____

# Lesson II Four-Square Vocabulary
## *Measure on a Number Line*

Write the definition for *number line*. Write what the
term means, draw a picture, and write your own
sentence using the term.

| **Definition** | **My Own Words** |
|---|---|
| A line with number labels. | See students' examples. |
| **My Picture** | **My Sentence** |
| See students' drawings. | A ruler is like a number line. |

( **number line** )

 **Teacher Directions:** Provide a description, explanation, or example of the new term
using images or real objects. Have students use the Glossary to write the definition.
Direct students to write a definition in their own words and draw a picture representing
their math term. Have students write a sentence using the term. Then encourage
students to read their sentence to a peer.

Grade 2 · Chapter II *Customary and Metric Lengths*  **101**

# Lesson 12 Measurement Data

## English Learner Instructional Strategy

### Collaborative Support: Round the Table

Place students into multilingual groups of 4 or 5. Distribute one problem from Exercises 7–9 to each group. Have students work jointly on the problem by passing the paper around the table. Direct each member of the group to write with a different color. Provide a step-by-step list for groups to follow, such as:

1) Read the problem aloud as a group and discuss.
2) One student underlines what they know.
3) The next student circles what they need to find.
4) The next student writes a plan.
5) The next student solves the problem with group input.
6) The last student checks for reasonableness.
7) Choose one student to present the solution to the class.

### English Language Development Leveled Activities

| Emerging Level | Expanding Level | Bridging Level |
|---|---|---|
| **Word Recognition** | **Sentence Frames** | **Report Back** |
| Have several volunteers come to the front. Measure each student's height in inches, and record the data in a line plot. Point to the numbers in the line plot, and say, *This is data. This data tells how tall these students are.* Ask questions about the data that students can answer with a gesture or one-word answer. For example, point to the Xs above a number, and ask, *How many students are ____ inches tall?* Have students count chorally. Point to a non-corresponding measurement and ask, *Is this the tallest student?* **no** | Divide students into two groups, and give a centimeter ruler and a line plot labeled 1 to 10 to each group. Say, *Use the ruler to measure each student's pointer finger.* Record the data on the line plot. Have the students use the following sentence frames to describe the data: **Most fingers were centimeters ____ long. The longest/shortest finger was ____ centimeters long. ____ students have fingers ____ centimeters long.** | Divide students into groups. Distribute a blank line plot to each group. Say, *Find 20 items in the room that you can measure and record in the line plot.* Direct students to measure using centimeters or inches. After groups have chosen objects to measure, circulate to help them label their line plots with an appropriate range of numbers. Have each group present a short report describing what they chose to measure and what the data on their line plot shows. |

### Teacher Notes:

NAME _____ DATE _____

# Lesson 12 Vocabulary Word Study
## *Measurement Data*

**Circle the correct word to complete the sentence.**

**I.** A line plot is a _____ that shows how often a number occurs in data.

ruler        (graph)

---

Show what you know about the term:
# line plot

There are __8__ letters.

There are __3__ vowels.

There are __5__ consonants.

__3__ vowels + __5__ consonants = __8__ letters in all.

---

**Draw a picture to show what *line plot* means.**

See students' examples.

**Teacher Directions:** Provide a description, explanation, or example of the new term using images or real objects. Read the sentence and have students circle the correct word. Direct students to count the letters, vowels and consonants in the math term, then complete the addition number sentence. Guide students to draw a picture representing their math term. Then encourage students to describe their picture to a peer.

# Chapter 12 Geometric Shapes and Equal Shares

## What's the Math in This Chapter?

### Mathematical Practice 3: Construct viable arguments and critique the reasoning of others

Draw 2 triangles, 1 circle, 1 square, and 1 rectangle on the board. Say, *I am going to circle the shapes that have four sides.* Circle the square and the circle. Ask, *Did I circle the correct shapes? Do both of my shapes have 4 sides?* Allow time for students to think. Then have them turn and talk with a peer. Regroup and discuss their observations. The discussion goal should be for students to "construct viable arguments" proving that you should not have circled the circle. You should have circled the rectangle instead.

Erase the circle around the circle shape and draw a circle around the rectangle. Ask, *Should I circle the triangles?* **no** Have students discuss why the triangles should not be circled. Explain to students how they critiqued your reasoning (circling the shapes that have 4 sides) and provided viable arguments as to why placing a circle around the circle shape was not correct. Say, *All of you were applying Mathematical Practice 3. You all explained your thinking and I understood what you meant.*

Display a chart with Mathematical Practice 3. Restate Mathematical Practice 3 and have students assist in rewriting it as an "I can" statement, for example: **I can explain my thinking so others can understand what I mean.** Post the new "I can" statement.

## Inquiry of the Essential Question:

### How do I use shapes and equal parts?

Inquiry Activity Target: **Students come to a conclusion that they must explain their reasoning when classifying shapes.**

As an introduction to the chapter, present the Essential Question to students. The inquiry graphic organizer will offer opportunities for students to observe, make inferences, and apply prior knowledge of geometric shapes representing the Essential Question. As they investigate, encourage students to draw, write, and collaborate with peers to demonstrate their observations and thinking. Then have students present additional questions they may have to a peer to extend discussions.

Regroup students and restate Mathematical Practice 3 and the Essential Question. Pose questions to reflect on what has been learned to guide students in making connections between the Mathematical Practice and the Essential Question.

NAME _____ DATE _____

# Chapter 12 Geometric Shapes and Equal Shares
## *Inquiry of the Essential Question:*

**How do I use shapes and equal parts?**

Read the Essential Question. Describe your observations
(I see…), inferences (I think…), and prior knowledge (I know…)
of each math example. Write additional questions you have
below. Then share your ideas and questions with a classmate.

four fourths

I see …

I think …

I know …

Circle the squares.

I see …

I think …

I know …

6 faces
12 edges
8 vertices

I see …

I think …

I know …

Questions I have…

_____

_____

 **Teacher Directions:** Read the Essential Question for students. Have students echo read.
Direct students to describe their observations, inferences, and prior knowledge of each
math example. Encourage students to write or draw additional questions they may have.
Then have students share their ideas/questions with a peer.

Grade 2 • **Chapter 12** *Geometric Shapes and Equal Shares* **103**

# Lesson 1 Two-Dimensional Shapes

## *English Learner Instructional Strategy*

### Vocabulary Support: Frontload Academic Vocabulary

Review the terms: *circle, rectangle, square and triangle.* Describe and model each shape with realia. Have students write a definition and draw a visual math example in their math journals.

Pair students. Assign each pair one of the two-dimensional shapes from this lesson and have them write a list or draw pictures of classroom objects that include their shape. Have each pair share their list with the group.

### English Language Development Leveled Activities

| Emerging Level | Expanding Level | Bridging Level |
|---|---|---|
| **Act It Out** | **Shape Game** | **Word Knowledge** |
| Write *triangle* on the board, and then show a model of a triangle. On chart paper, draw a large triangle. Set the drawing on the ground, and invite three students to stand around the triangle, one student per side. Count the students, and ask, *How many students are around the triangle?* Have students answer **3** chorally or with a gesture. Then ask, *How many sides are on a triangle?* Have students answer **3** chorally or with a gesture. Write 3 next to the triangle on the board. Repeat the activity with a rectangle, pentagon, and hexagon. | Divide students into four groups. Write *triangle, rectangle, pentagon,* and *hexagon* on four index cards. Give each group a card. Have groups use gestures or manipulatives to demonstrate their shape, and have the other groups guess the shape's name. If the guess is correct, the demonstration group shows a thumbs-up, if the guess is incorrect they show a thumbs-down. | Say a shape name. Have students find an example in the classroom and describe its characteristics. Repeat the activity by describing the characteristics of a shape. Then have students find an example in the classroom and identify the shape's name. Finally, have one student describe a shape they see in the classroom, and have a second student identify the shape by name. |

### Teacher Notes:

NAME _____ DATE _____

# Lesson I Vocabulary Definition Map
## *Two-Dimensional Shapes*

Use the definition map to write what the math word means and tell what the word is like. Write or draw a math example and share it with a classmate.

**My Math Word:**

two-dimensional shapes

**What It Is Like:**

A ___circle___ is closed and round.

A ___triangle___ has 3 sides and 3 angles.

**What It Means:**

The outline of a shape that has only length and width.

Shapes with 4 sides are the ___square___, ___rectangle___, ___parallelogram___, and ___trapezoid___.

A ___pentagon___ has 5 sides and 5 angles.

A ___hexagon___ has 6 sides and 6 angles.

**My Math Example:**

See students' examples.

**Teacher Directions:** Provide a description, explanation, or example of the new term using images or real objects. Have students use the lesson or Glossary to define the math term. Direct students to list characteristics, and draw a picture representing their math term. Then encourage students to describe their picture to a peer.

# Lesson 2 Sides and Angles

## English Learner Instructional Strategy

### Language Structure Support: Communication Guide

Have students use the glossary to review the new vocabulary in both English and Spanish for this lesson. Have students write definitions and draw math examples for each term in their math journals. Provide an appropriate translation tool for non-Spanish speaking ELs. Post and model using the following communication guide for students to utilize during the shape lessons.

This is a _____ (triangle, quadrilateral, pentagon, hexagon, circle).
A _____ has _____ sides. A _____ has _____ angles.
This shape is a _____ because it has _____ sides.

### English Language Development Leveled Activities

| Emerging Level | Expanding Level | Bridging Level |
|---|---|---|
| **Number Recognition** | **Identify and Describe** | **Written Language** |
| Draw a triangle on the board, and label each of the 3 angles as 1, 2, and 3. Say, *An angle is where two sides meet. I will clap for each angle.* Clap as you count each angle. Say, *Clap for each angle.* Have students clap with you as you count the angles a second time. Ask, *How many angles?* **3** Display examples of other shapes in random order. Have students clap to show how many angles they see in each shape. For example, show a pentagon. Students will clap five times. Then say, *A pentagon has five angles.* Ask, *How many angles?* **5** | Divide students into groups, and distribute a model of a different shape to each group. Have groups work together to identify the name of their shape and the number of sides and angles it has. Have a volunteer from each group describe their shape to the class using the sentence frame: **A _____ has _____ sides and _____ angles.** | On the board, write: *more, most, fewer, fewest,* and *same.* Have students work in pairs. Assign three different shapes to each pair. (Shapes should vary by pair.) Say, *Write three sentences comparing your shapes. Use one of the words on the board in each of your sentences.* Have students share their comparisons. For example, a pair with a circle, pentagon, and triangle might write: **A triangle has two fewer sides than a pentagon. A circle has the fewest number of angles. A pentagon has the same number of sides as angles.** |

**Teacher Notes:**

NAME _____ DATE _____

# Lesson 2 Note Taking
## *Sides and Angles*

Read the question. Write words you need help with. Use your lesson to write your Cornell notes.

| **Building on the Essential Question** | **Notes:** |
|---|---|
| How can I use sides and angles to describe two-dimensional shapes? | A two-dimensional shape is the outline of a shape that has only <u>length</u> and <u>width</u>. |

A two-dimensional shape is the outline of a shape that has only <u>length</u> and <u>width</u>.

One of the line segments that make up a shape is a <u>side</u>.

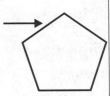

Two sides on a two-dimensional shape meet to form an <u>angle</u>.

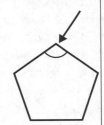

A <u>circle</u> has 0 sides and 0 angles.

A <u>triangle</u> has 3 sides and 3 angles.

A <u>quadrilateral</u> has 4 sides and 4 angles.

A <u>pentagon</u> has 5 sides and 5 angles.

A <u>hexagon</u> has 6 sides and 6 angles.

**Words I need help with:**
See students' words.

**Teacher Directions:** Read the Building on the Essential Question and have students list words/phrases they need assistance with. Provide descriptions, explanations, or examples of the terms using images or real objects. Read each sentence frame and have students write the appropriate terms. Have students read their notes aloud.

Grade 2 • **Chapter 12** *Geometric Shapes and Equal Shares* **105**

# Lesson 3 Problem Solving Strategy: Draw a Diagram

## English Learner Instructional Strategy

### Collaborative Support: Round the Table

For extended practice, place students into multilingual groups of 4 or 5 students. Assign the Lesson 3 Reteach worksheet exercises to each group. Have students work jointly on the problems by passing the paper around the table for each member to provide input. Direct each member of the group to write with a different color to ensure all students participate. Provide a step-by-step list for groups to follow, such as: 1) Read the problem aloud as a group and discuss. 2) One student underlines what they know. 3) The next student circles what they need to find. 4) The next student highlights signal words and phrases. 5) The next student writes a plan. 6) The next student solves the problem. 7) The last student checks for reasonableness. 8) Choose one student to report back.

### English Language Development Leveled Activities

| Emerging Level | Expanding Level | Bridging Level |
|---|---|---|
| **Signal Words** | **Communication Guide** | **Public Speaking Norms** |
| On the board, write Understand, Plan, Solve, and Check. Then write and read aloud, *Stella drew a shape. The shape has four sides that are the same length. How many angles does the shape have?* Underline the first two sentences. Point to the question mark at the end of the of the last sentence, and say, *The sentence with a question mark tells us what we need to find.* Have students choral read the last sentence with you. Then point to Plan, and say, *Now we need a plan. We can* draw a diagram. Continue in this manner as you model solving the problem. | Draw four large boxes on the board, and label them Understand, Plan, Solve, and Check. In the Understand box, write then read aloud, *A shape has more angles than a square. It has fewer angles than a hexagon. What is the shape?* Provide the following sentence frames to help students guide you in completing the problem-solving graphic organizer: I **know _____, so I will underline _____. I need to find _____, so I will circle _____. I will _____ to find the answer. I will _____ to check the answer.** | Draw a large problem-solving graphic organizer on the board with four large boxes labeled Understand, Plan, Solve, and Check. Divide students into four groups, and confidentially give each group a different two-dimensional shape without letting the other groups see the shape. Say, *Describe your shape without naming it.* In the Understand box on the board, record the first group's description. Have the other students direct you in completing the graphic organizer to determine the shape. Repeat with each group's description. |

### Multicultural Teacher Tip

As the metric system is the standard throughout most parts of the world, ELs will most likely be more familiar with units of metric measurement than they will be with standard units. Students who have worked only with the metric system in the past will be more familiar with partial amounts written as decimals, not fractions.

NAME _____ DATE _____

# Lesson 3 Problem Solving
## STRATEGY: Draw a Diagram

<u>Underline</u> what you know. (Circle) what you need
to find. Draw a diagram to solve the problems.

**I.** <u>If a shape has 3 sides and 3 angles,</u>

(what shape is it?)

Draw the shape.

The shape is a ___triangle___ .

---

**2. Abby draws** <u>a triangle.</u>

**Samuel draws** <u>a shape that has I more</u>
<u>side than a triangle.</u>

(What shape **did** Samuel **draw**?)

Draw the shape.

Samuel drew a ___quadrilateral___ .

triangle

**Teacher Directions:** Provide a description, explanation, or example of the boldface
terms and nouns using images or real objects. Read each sentence and have students
echo read. Encourage students to draw a diagram and then write their answers in each
restated question. Have students read each answer sentence aloud.

# Lesson 4 Three-Dimensional Shapes

## English Learner Instructional Strategy

### Vocabulary Support: Vocabulary Game

Write the following words on index cards: *cone, cube, cylinder, pyramid, rectangular prism,* and *sphere*. Draw an example of each on cards. Create a set for each pair of students. Demonstrate how to play the "memory" game. Shuffle the cards and place facedown in rows. Players take turns flipping over two cards. If the cards create a match (show a shape and matching name), they are set aside. If the cards are not a match, they are flipped facedown. Play until all cards are matched.

### English Language Development Leveled Activities

| Emerging Level | Expanding Level | Bridging Level |
|---|---|---|
| **Phonemic Awareness** | **Word Identification** | **Shape Game** |
| Display a cube. Say, *This is a cube.* Then pass it to a student, and say, You have a cube. Emphasize the /ü/ sound. Say, *Now you pass the cube.* Have the student hand the cube to the next student. Say, *Who has a cube?* Encourage students to say the cube holder's name. *Do you have a cube?* Point to the cube holder, and have the student answer orally or with a gesture, **yes**. Repeat with the other three-dimensional shapes. | Divide students into five groups. Assign each group as: Cone, Cube, Cylinder, Pyramid, or Rectangular Prism. Say, *Look around the classroom, and find as many examples of your assigned shape as you can.* Give students a few minutes to locate objects. Then have a volunteer from each group tell what objects they found. Provide a sentence frame, such as: **The _____ is shaped like a _____.** | Divide students into six groups. Confidentially give each group a cone, cube, cylinder, pyramid, rectangular prism, or sphere, without letting the other groups see which shape was given. Have groups take turns describing one or two common objects that have their assigned shape. For example, the group assigned to a sphere might say, **A basketball and a globe have this shape.** Have groups guess the shape being described. |

**Teacher Notes:**

NAME _____ DATE _____

# Lesson 4 Word Web

## *Three-Dimensional Shapes*

Use the word web to draw 6 examples of three-dimensional shapes.

Answers may be in any order.

See students' drawings.

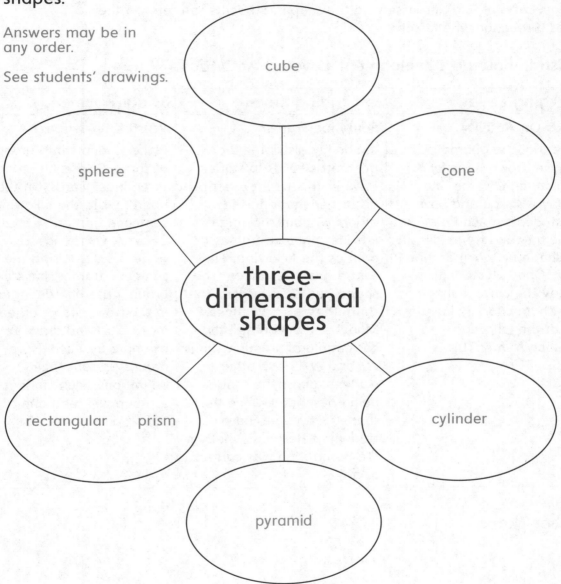

**Teacher Directions:** Review the term *three-dimensional shapes*. Review the names of each three-dimensional shape using images and manipulatives. Have students individually complete the web drawing each shape. Then have partners share their drawings using the sentence frame: **This three-dimensional shape is a _____.**

Grade 2 • Chapter 12 *Geometric Shapes and Equal Shares* **107**

# Lesson 5 Faces, Edges, and Vertices

## English Learner Instructional Strategy

### Sensory Support: Hands-On Activity

Provide students with the following Manipulative Masters worksheets: Square and Triangle Pyramid Patterns, Cube and Rectangular Prism Patterns, Cone and Cylinder Patterns. Use the net patterns to help students visualize, count, and label the number of faces, edges, and vertices on each shape. Then have students cut out, fold and assemble the shapes. Students can refer to their shapes throughout the lesson.

### English Language Development Leveled Activities

| Emerging Level | Expanding Level | Bridging Level |
|---|---|---|
| **Hands-On Activity** | **Sentence Frames** | **Word Knowledge** |
| Place 6 of the geometric solids in brown bags for each student. Name and display a shape and have the students reach inside of the bag and try to identify the shape by touch and pull it out. Once all students display the correct shape, have them chorally identify it with the following sentence frame: **This is a ____.** | Randomly distribute geometric solids to students. Have each student describe his or her shape to the others without showing it. Provide sentence frames, such as the following: **The shape has ____ faces. The shape has ____ edges. The shape has ____ vertices.** Model and encourage the correct plural pronunciations /s/ and /z/. Have other students guess the shape being described. Once the shape is revealed, have students determine whether the description was correct or not. | Place students into groups of three. Give each group three index cards. On the board, write the following sentence frames: **A face is ____. A vertex is ____. An edge is ____.** Using the sentence frames, have each group write one definition in their own words on each index card and draw an example for each. Have groups present their definitions. Each student in a group will read one of the definitions. |

**Teacher Notes:**

NAME _____ DATE _____

# Lesson 5 Word Identification

## *Faces, Edges, and Vertices*

Use the word bank to label the cube.

**Word Bank**

edge

face

vertex

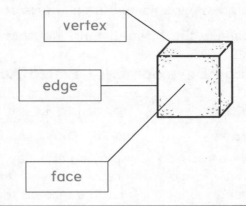

Write the correct term from above for each sentence on the blank lines.

**1.** A __face__ is the flat part of a three-dimensional shape.

**2.** An __edge__ is the line segment where 2 faces meet.

**3.** The corner of a box is a __vertex__.

**Teacher Directions:** Review the terms using images or real objects. Have students say each term and then draw a line to match the term to its meaning. Direct students to say each sentence, then write the corresponding term in the sentences. Encourage students to read the sentences to a peer.

# Lesson 6 Relate Shapes and Solids
## *English Learner Instructional Strategy*

### Sensory Support: Models/Drawings

Give each student the 6 different geometric solids. Have students work in pairs and use their math journals to describe the shape of the faces that make up each geometric solid using the following sentence frames:

**A cone/cube/cylinder/pyramid/rectangular prism/sphere has ____ faces.**

**The face(s) that make the three-dimensional shape are/is ____.**

### English Language Development Leveled Activities

| Emerging Level | Expanding Level | Bridging Level |
|---|---|---|
| **Making Connections** | **Building Oral Language** | **Sentence Frames** |
| Give each student a two-dimensional triangle, circle, square, and rectangle. Show a pyramid. Point to one face, and say, *This face has a ____ shape.* Lay the face against the board or chart paper and trace the shape. Say, *What is this shape? Show me the shape.* Have students answer by displaying their triangles. Repeat this process for each face of the pyramid. When all five faces have been traced, say, *The pyramid has five faces. Four faces are triangles. One face is a square.* Repeat the activity with the other three-dimensional shapes. | Display a cube. Say, *I will trace one face of this shape.* Trace the face on the board or chart paper. Say, *The shape I traced is a square. I will trace the other faces.* Finish tracing each face, and say, *I traced all six faces of the cube. They are all squares.* Give each student a three-dimensional shape. Say, *Trace each face of your shape.* After students have traced each face, have them describe what they traced using the sentence frames: **I traced the faces of a ____. I traced ____ triangles/circles/square(s)/rectangles.** | Have students work in pairs. Randomly assign each pair a three-dimensional shape. Say, *Find an object in the classroom that has the same shape as your three-dimensional shape. Explain how you know it is that shape.* Give students a few minutes to find their objects. Then have each pair explain how they know the object is the shape they were assigned. Ensure students use the following terms as they report back: *three-dimensional, faces, sides, vertices.* |

**Teacher Notes:**

NAME _____ DATE _____

# Lesson 6 Multiple Meaning Word
## *Relate Shapes and Solids*

Spell then say the math word. Draw a picture that shows the math word meaning and the non-math word meaning in the boxes.

| face | |
|---|---|
| **Math Meaning**<br><br>Students' examples should represent the flat part of a three-dimensional shape. | **Non-Math Meaning**<br><br>See students' examples. |

Use the sentence frame below to help you describe your pictures.

This picture for the word ___face___ shows _____.

**Teacher Directions:** Provide math and non-math descriptions, explanations, or examples of the new term using images or real objects. Have students spell then say the term. Then direct students to draw pictures showing a math and non-math meaning of the math term. Encourage students to describe their pictures to a peer using the sentence frame.

Grade 2 • **Chapter 12** *Geometric Shapes and Equal Shares* **109**

# Lesson 7 Halves, Thirds, and Fourths

## English Learner Instructional Strategy

### Collaborative Support: Echo Reading/Mentors

Write *halves, thirds,* and *fourths* on the Math Word Wall. Have students refer to the glossary and their My Vocabulary Cards to write a definition and draw a visual math example in their math journals.

Pair ELs with a native English-speaking student during the Problem Solving Exercises 17–19. Have the native English-speaking student read the problem, and then have the EL echo read. Encourage the EL to ask, **What's this word?** when they are unsure of a word's meaning. Direct English-speaking mentors to be creative when modeling meanings of words.

### English Language Development Leveled Activities

| Emerging Level | Expanding Level | Bridging Level |
|---|---|---|
| **Academic Word Knowledge** | **Choral Responses** | **Report Back** |
| Write *half* on the left side of the board and again on the right side of the board. Invite an even number of students to the board. Indicate the whole group as you say, *This is one group of students.* Guide half of the students to stand by the left side of the board and the other half on the right side. Point to each word on the board as you say, *Half of the group is here, and half of the group is here. How many halves?* **2** Emphasize the difference between *half* and *halves* as you say the terms. Repeat with *thirds* and *fourths.* | Invite an even number of students to the front, then divide them into two equal groups. Say, *I divided the group into two halves, or two equal parts.* Point to one half, and say, *Now this half will laugh.* Have the students laugh. Repeat, having the other half cry. Have the class repeat chorally, **There are two halves. One half laughed. The other half cried.** Repeat the activity to demonstrate *third* and *thirds,* as well as *fourth* and *fourths.* | Divide students into three groups, and give each student a paper shape. Assign each group as either Halves, Thirds, or Fourths. Say, *Fold, then draw lines on your shape to show the equal parts your group was assigned.* Encourage groups to discuss how to partition each shape. Then have each group report back using the terms: *equal parts, halves, thirds, fourths.* Be sure students use the /z/ or /s/ sound correctly to indicate plurals. |

**Teacher Notes:**

NAME _____ DATE _____

# Lesson 7 Guided Writing
## *Halves, Thirds, and Fourths*

**How do you find halves, thirds, and fourths?**

Use the exercises below to help you build on answering the Essential Question. Write the correct word or phrase on the lines provided.

I.  What key words do you see in the question?
    halves, thirds, fourths

2.  What is another word for *separate* or *break apart*?
    partition

3.  If I partition a whole into two __equal__ parts, I have two
    __halves__ .

4.  Three equal parts of a whole is three __thirds__ .

5.  Four equal parts is four __fourths__ .

6.  How do you find halves, thirds, and fourths?

    I can break apart a whole into two equal parts to make halves, or three

    equal parts to make thirds, or four equal parts to make fourths.

**Teacher Directions:** Read the Building on the Essential Question. Provide descriptions, explanations, or examples of the terms using images or real objects. Read each sentence frame and have students use the lesson and Glossary to write the appropriate terms. Have students read the sentences aloud.

# Lesson 8 Area
## English Learner Instructional Strategy

### Sensory Support: Pictures and Realia

Review English terms from the Problem Solving exercises which may be unfamiliar to students, such as: *dominoes, crackers, tray/plate, row,* and *column* to provide additional language support. Use pictures, realia, and modeling to clarify the terms. Pair emerging/expanding students with same-native language bridging level students during the exercises. Allow use of native language to clarify meaning.

### English Language Development Leveled Activities

| Emerging Level | Expanding Level | Bridging Level |
|---|---|---|
| **Build Background Knowledge** <br><br> Distribute rectangular sheets of paper to each student. Model folding the paper vertically in half and have students mimic. Say, *One rectangle, two* **equal parts**. Fold a horizontal line through the rectangle so it is divided into four parts. Have students mimic. Ask, *How many equal parts now?* Give students a moment to answer chorally or with a gesture, **4 equal parts.** Ask, *How many parts if we fold it again?* Have students refold on previous folds and then fold in half again. Have students unfold and count the equal parts. Ask, *How many equal parts?* **8 equal parts** | **Listen and Identify** <br><br> Pair students and distribute a sheet of paper to each pair. Count by twos to assign each pair a number. Say, *Draw a rectangle. Then partition your rectangle into equal-sized squares. The number of squares will be the number you were assigned.* Give pairs a few minutes to complete the task. Then have one student from each pair describe what they did using the sentence frame: **We partitioned our rectangle into ____ equal-sized squares.** Be sure students are correctly saying the /d/ sound at the end of *partitioned.* | **Shape Game** <br><br> Pair students. Distribute 20 color tiles to each student. Student A will create a rectangle from their set of tiles. For example, 12 tiles could be arranged into a 3 by 4 array. Student B will recreate the same rectangle using the appropriate number of tiles. Students will count the tiles in both rectangles to verify that the students used the same number of tiles. Have students switch roles and repeat the activity. |

**Teacher Notes:**

NAME _____  DATE _____

# Lesson 8 Four-Square Vocabulary
## *Area*

Write the definition for *partition*. Write what the
word means, draw a picture, and write your own
sentence using the word.

| **Definition** | **My Own Words** |
|---|---|
| To divide or "break up." | See students' examples. |
| **My Picture** | **My Sentence** |
| See students' drawings. | Sample sentence: I can partition the pizza into 8 slices. |

partition

**Teacher Directions:** Provide a description, explanation, or example of the new term
using images or real objects. Have students use the Glossary to write the definition.
Direct students to write a definition in their own words and draw a picture representing
their math term. Have students write a sentence using the term. Then encourage
students to read their sentence to a peer.

Grade 2 • **Chapter 12** *Geometric Shapes and Equal Shares*    III

An Interview with
# Dinah Zike Explaining
# Visual Kinesthetic Vocabulary®, or VKVs®

## What are VKVs and who needs them?

❝ VKVs are flashcards that animate words by kinesthetically focusing on their structure, use, and meaning. VKVs are beneficial not only to students learning the specialized vocabulary of a content area, but also to students learning the vocabulary of a second language. ❞

**Dinah Zike | Educational Consultant**

Dinah-Might Activities, Inc. – San Antonio, Texas

## Why did you invent VKVs?

❝ Twenty years ago, I began designing flashcards that would accomplish the same thing with academic vocabulary and cognates that Foldables® do with general information, concepts, and ideas—make them a visual, kinesthetic, and memorable experience. ❞

**I had three goals in mind:**

- **Making two-dimensional flashcards three-dimensional**

- **Designing flashcards that allow one or more parts of a word or phrase to be manipulated and changed to form numerous terms based upon a commonality**

- **Using one sheet or strip of paper to make purposefully shaped flashcards that were neither glued nor stapled, but could be folded to the same height, making them easy to stack and store**

## Why are VKVs important in today's classroom?

❝ At the beginning of this century, research and reports indicated the importance of vocabulary to overall academic achievement. This research resulted in a more comprehensive teaching of academic vocabulary and a focus on the use of cognates to help students learn a second language. Teachers know the importance of using a variety of strategies to teach vocabulary to a diverse population of students. VKVs function as one of those strategies. ❞

Dinah Zike's
**Visual Kinesthetic Vocabulary**®

## How are VKVs used to teach content vocabulary to EL students?

" VKVs can be used to show the similarities between cognates in Spanish and English. For example, by folding and unfolding specially designed VKVs, students can experience English terms in one color and Spanish in a second color on the same flashcard while noting the similarities in their roots. "

## What organization and usage hints would you give teachers using VKVs?

" Cut off the flap of a 6" x 9" envelope and slightly widen the envelope's opening by cutting away a shallow V or half circle on one side only. Glue the non-cut side of the envelope into the front or back of student workbooks or journals. VKVs can be stored in the pocket.

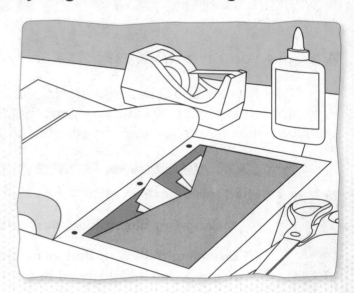

Encourage students to individualize their flashcards by writing notes, sketching diagrams, recording examples, forming plurals (radius: radii or radiuses), and noting when the math terms presented are homophones (sine/sign) or contain root words or combining forms (kilo-, milli-, tri-).

As students make and use the flashcards included in this text, they will learn how to design their own VKVs. Provide time for students to design, create, and share their flashcards with classmates. "

Dinah Zike's book Foldables, Notebook Foldables, & VKVs for Spelling and Vocabulary 4th-12th won a Teachers' Choice Award in 2011 for "instructional value, ease of use, quality, and innovation"; it has become a popular methods resource for teaching and learning vocabulary.

Dinah Zike's
Visual
Kinesthetic
Vocabulary®

VKV

Chapter I

✂ cut on all dashed lines          fold on all solid lines

Write a number sentence with the addends 1 and 3. Circle the sum. (Escribe un enunciado numérico con los sumandos 1 y 3. Encierra en un círculo la suma.)

____ + ____ = ____

Write math symbols to complete each subtraction sentence. (Escribe símbolos matemáticos para completar los enunciados de resta.)

8 ◯ 2 ◯ 6
5 ◯ 0 ◯ 5
8 ◯ 4 ◯ 4

subtract

sum

Write math symbols to complete each addition sentence. (Escribe símbolos matemáticos para completar los enunciados de suma.)

5 ◯ 3 ◯ 8
4 ◯ 1 ◯ 5
2 ◯ 7 ◯ 9

add

Dinah Zike's
Visual
Kinesthetic
Vocabulary®

Chapter I

✂ cut on all dashed lines

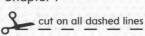 fold on all solid lines

a

restar

sumar

Subtract to find each difference. (Halla cuánto da cada diferencia.)

5 − 2 = _____    8 − 0 = _____

7 − 3 = _____    7 − 1 = _____

Add to find each sum. (Halla cuánto da cada suma.)

2 + 5 = _____    8 + 0 = _____

3 + 6 = _____    6 + 1 = _____

Find each sum. (Halla las sumas.)

5 + 2 = _____    3 + 1 = _____

4 + 5 = _____    6 + 2 = _____

**difference**

**doubles**

**sum**

The difference is the answer in a _____ problem.

(La diferencia es la respuesta en un problema de _____.)

The sum is the answer in an _____ problem.

(La suma es la respuesta en un problema de _____.)

Add using doubles facts.
(Suma usando sumas de dobles.)

$6 + 6 =$ _____

$4 + 4 =$ _____

$8 + 8 =$ _____

# diferencia

# suma

Lacey gave 4 beads to Maddie. She gave 5 beads to Max. How many total beads did Lacey give away? (Lacey le dio 4 cuentas a Maddie. Le dio 5 cuentas a Max. ¿Cuántas cuentas en total dio Lacey?)

_____ beads (cuentas)

Did you use a doubles fact or a near doubles fact to solve? (¿Usaste una suma de dobles o una suma de casi dobles para resolver el problema?) _____

Find each difference. (Halla las diferencias.)

12 − 2 = _____     14 − 7 = _____

9 − 3 = _____     10 − 4 = _____

Find each sum. (Halla las sumas.)

4 + 5 = _____     9 + 4 = _____

8 + 3 = _____     6 + 6 = _____

do

count on

count back

Use the number line to count back to subtract. (Usa la recta numérica para contar hacia atrás y restar.)

← 0 1 2 3 4 5 6 7 8 9 10 11 12 →

9 − 3 = ___         10 − 8 = ___

11 − 4 = ___         6 − 3 = ___

Use the number line to count on to add. (Usa la recta numérica para seguir contando y sumar.)

← 0 1 2 3 4 5 6 7 8 9 10 11 12 →

5 + 4 = ___         7 + 3 = ___

3 + 9 = ___         6 + 4 = ___

Dinah Zike's
**VKV** Visual
Kinesthetic
Vocabulary®

Chapter I

✂ cut on all dashed lines     fold on all solid lines

seguir contando

contar hacia atrás

10 birds sit on a wire. 2 birds fly away. How many birds are left? Count back to subtract. (10 pájaros están sobre un cable. 2 pájaros vuelan. ¿Cuántos pájaros quedan? Cuenta hacia atrás para restar.)

—— birds (pájaros)

A tree branch holds 6 apples. Another branch holds 5 apples. How many apples on both branches? Count on to add. (La rama de un árbol tiene 6 manzanas. Otra rama tiene 5 manzanas. ¿Cuántas manzanas hay en ambas ramas? Sigue contando para sumar.)

—— apples (manzanas)

odd number

par

regroup

You must regroup when (Debes reagrupar cuando) _____

Write three even numbers. (Escribe tres números pares.)

Write three odd numbers. (Escribe tres números impares.)

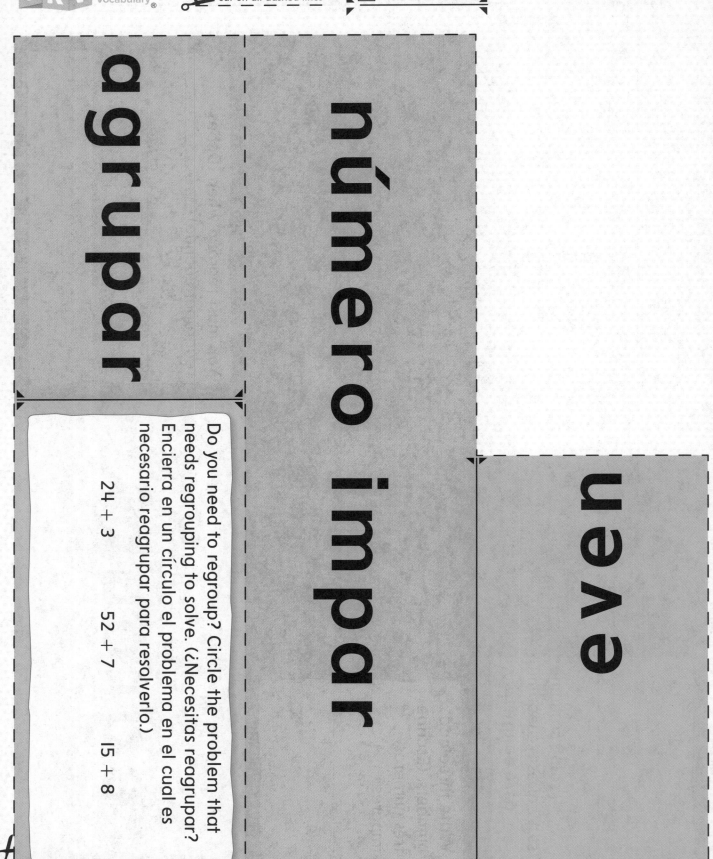

agrupar

número impar

even

Do you need to regroup? Circle the problem that needs regrouping to solve. (¿Necesitas reagrupar? Encierra en un círculo el problema en el cual es necesario reagrupar para resolverlo.)

24 + 3     52 + 7     15 + 8

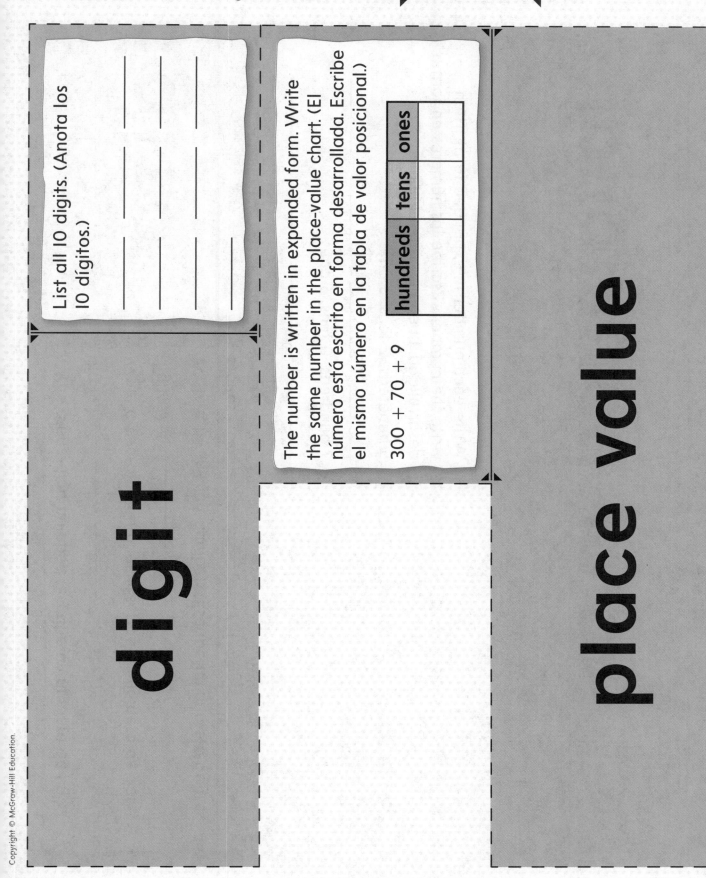

List all 10 digits. (Anota los 10 dígitos.)

The number is written in expanded form. Write the same number in the place-value chart. (El número está escrito en forma desarrollada. Escribe el mismo número en la tabla de valor posicional.)

| hundreds | tens | ones |
|----------|------|------|
|          |      |      |

300 + 70 + 9

digit

place value

# valor posicional

Write each number in expanded form. Then write the number. (Escribe los números en forma desarrollada. Luego escribe el número.)

5 hundreds, 2 tens, 3 ones

___ + ___ + ___ = ___

7 hundreds, 0 tens, 6 ones

___ + ___ + ___ = ___

# ígito

Write a one-digit number. (Escribe un número de un dígito.)

___

Write a two-digit number. (Escribe un número de dos dígitos.)

___

Write a three-digit number. (Escribe un número de tres dígitos.)

___

Circle the digits in the thousands place. (Encierra en un círculo los dígitos en la posición de los millares.)

527        1,415

2,365      13

48         942

## thousands

Circle the correct word. When you compare two numbers, start with digits in the (greatest, least) place value. (Encierra en un círculo la palabra correcta. Cuando comparas dos números, comienzas con los dígitos en el (mayor, menor) valor de posición.)

Circle the digits in the hundreds place. (Encierra en un círculo los dígitos en la posición de las centenas.)

1,527      844

25         1,013

649        79

## hundreds

## compare

**ar**

**millares**

**centenas**

Read the number. Write the number. (Lee el número. Escribe el número.)

one thousand, four hundred twelve (mil cuatrocientos doce) _____

one thousand, one hundred twenty-three (mil ciento veintitrés) _____

Read the number. Write the number. (Lee el número. Escribe el número.)

four hundred sixteen (cuatrocientos dieciséis) _____

one hundred thirty-two (ciento treinta y dos) _____

seven hundred eighty-five (setecientos ochenta y cinco) _____

Compare. Write <, >, or =. (Compara. Escribe <, > o =.)

115 ◯ 248        845 ◯ 845

647 ◯ 321        745 ◯ 796

355 ◯ 255        413 ◯ 411

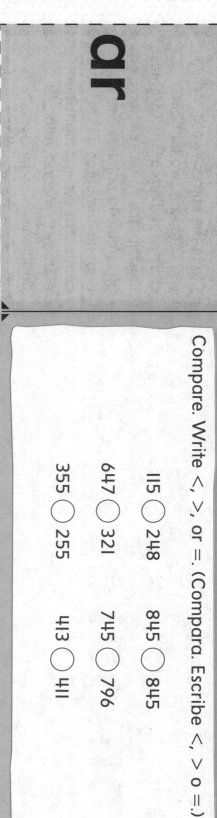

Dinah Zike's
Visual
Kinesthetic
Vocabulary.®

✂ cut on all dashed lines          🔲 fold on all solid lines

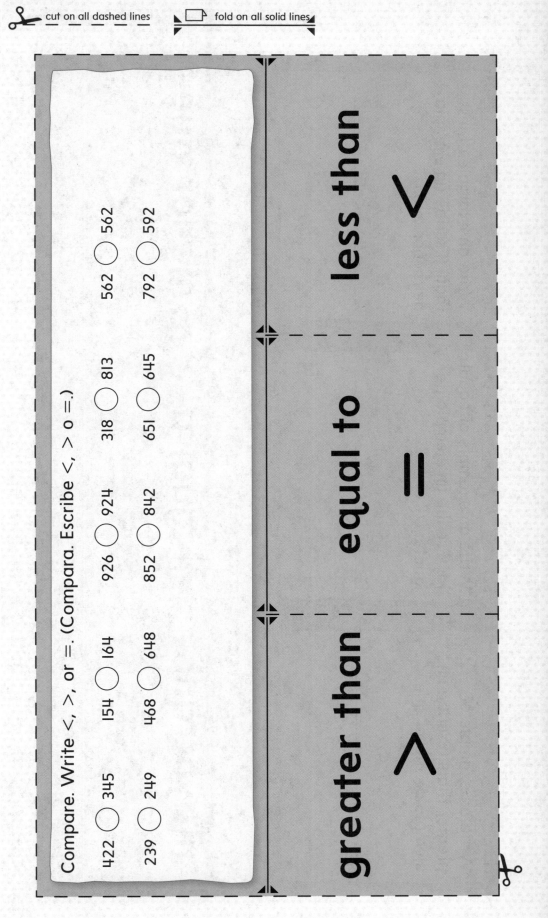

less than

<

equal to

=

greater than

>

Compare. Write <, >, or =. (Compara. Escribe <, > o =.)

422 ◯ 345        926 ◯ 924        562 ◯ 562

239 ◯ 249        852 ◯ 842        792 ◯ 592

154 ◯ 164        318 ◯ 813

468 ◯ 648        651 ◯ 645

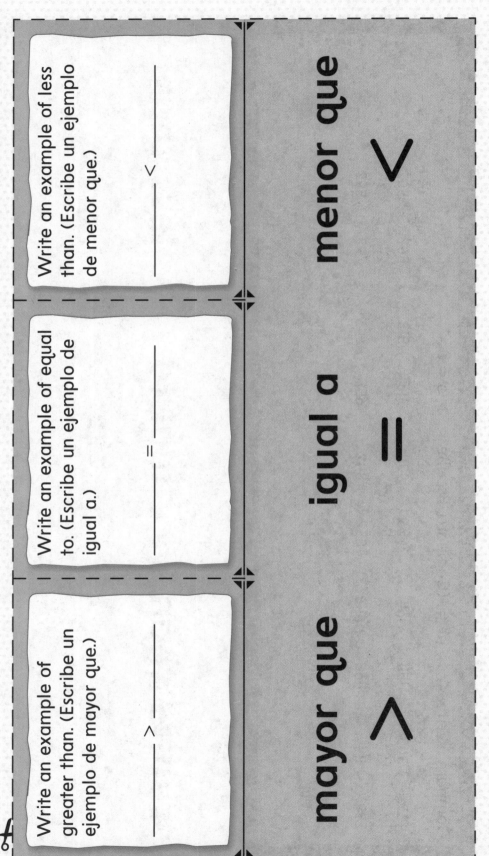

Write an example of less than. (Escribe un ejemplo de menor que.)

<

menor que

>

Write an example of equal to. (Escribe un ejemplo de igual a.)

=

igual a

=

Write an example of greater than. (Escribe un ejemplo de mayor que.)

>

mayor que

>

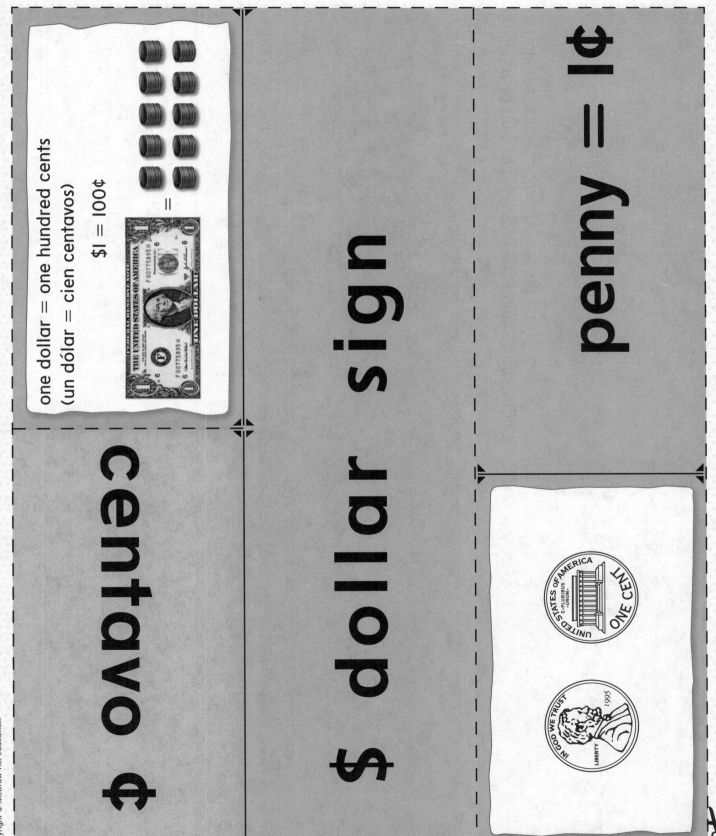

one dollar = one hundred cents
(un dólar = cien centavos)

$1 = 100¢

penny = 1¢

centavo ¢

$ dollar sign

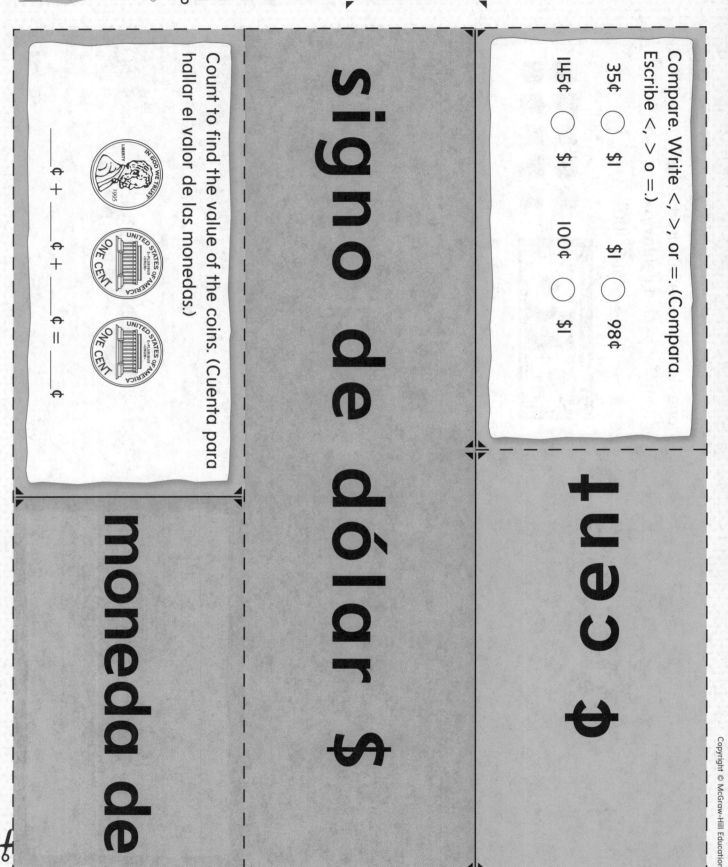

Compare. Write <, >, or =. (Compara.
Escribe <, > o =.)

35¢ ◯ $1

145¢ ◯ $1

$1 ◯ 98¢

100¢ ◯ $1

Count to find the value of the coins. (Cuenta para
hallar el valor de las monedas.)

_____ ¢ + _____ ¢ + _____ ¢ = _____ ¢

signo de dólar $

¢ cent

moneda de

# VKV
Dinah Zike's
**Visual Kinesthetic Vocabulary**®

Chapter 8

✂ cut on all dashed lines

☐ fold on all solid lines

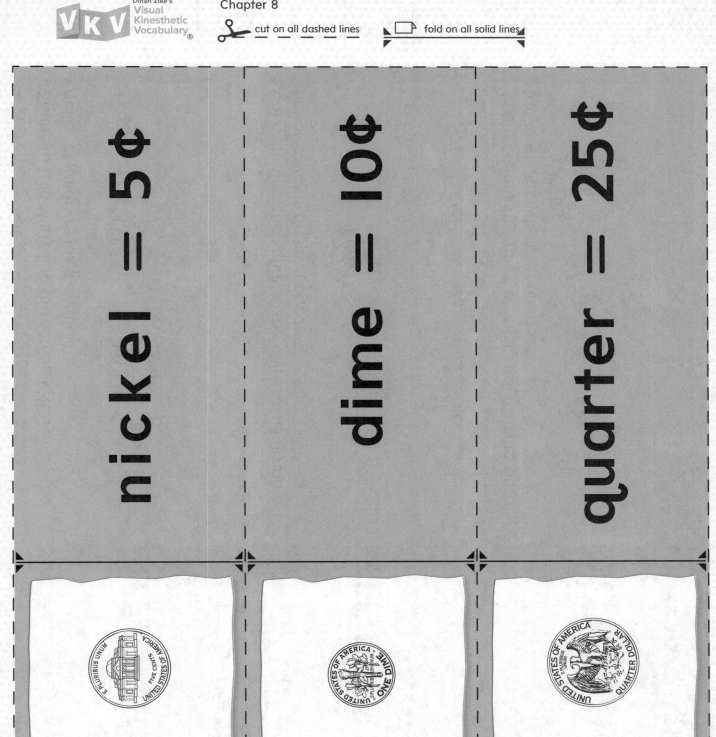

nickel = 5¢

dime = 10¢

quarter = 25¢

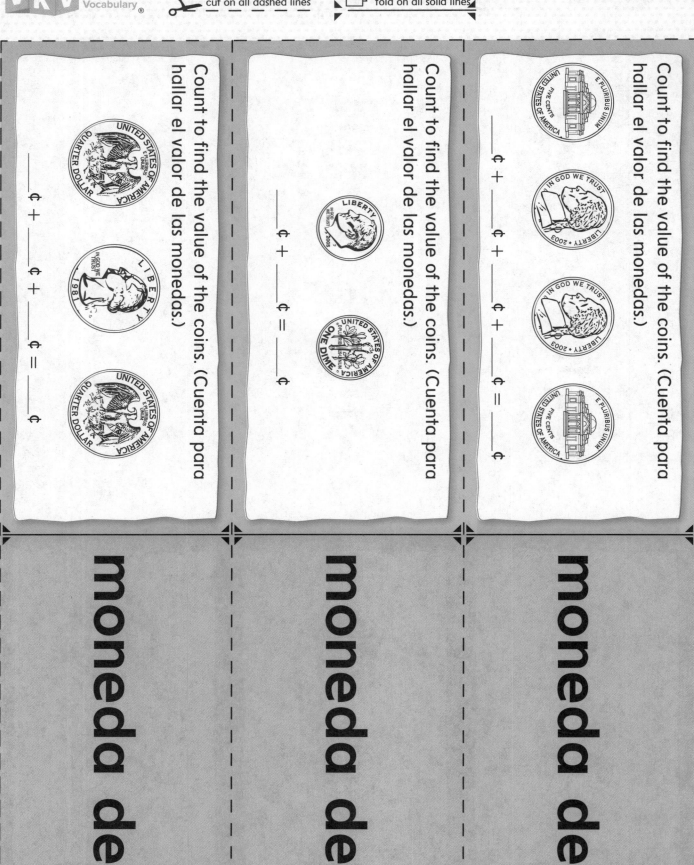

Count to find the value of the coins. (Cuenta para hallar el valor de las monedas.)

¢ + ¢ + ¢ = ¢

Count to find the value of the coins. (Cuenta para hallar el valor de las monedas.)

¢ + ¢ = ¢

Count to find the value of the coins. (Cuenta para hallar el valor de las monedas.)

¢ + ¢ + ¢ + ¢ = ¢

**moneda de**

**moneda de**

**moneda de**

Dinah Zike's
**VKV** Visual
Kinesthetic
Vocabulary ®

Chapter 9

✂ cut on all dashed lines

fold on all solid lines

Write the total for each choice. (Escribe el total para cada opción.)

| Favorite Shoes | Tally | Total |
|---|---|---|
| Tennis Shoes | ⊬⊬⊬ | |
| Flip Flops | ||| | |
| Boots | || | |

**picture graph**

What activity does each symbol represent? Write your ideas on the graph. (¿Qué actividad representa cada símbolo? Escribe tus ideas en la gráfica.)

**Favorite Summer Activity**

Key: Each picture = 1 vote

**symbol**

Data is (Los datos son)

**data**

Dinah Zike's
**VKV**
Visual
Kinesthetic
Vocabulary®

Chapter 9

✂ cut on all dashed lines

fold on all solid lines

# gráfica con imágenes

# ímbolo

# os

Use the tally chart from the other side to make a picture graph. (Usa la tabla de conteo del otro lado para hacer una gráfica con imágenes.)

**Favorite Shoes**

| | | | | | |
|---|---|---|---|---|---|
| Tennis Shoes | | | | | |
| Flip Flops | | | | | |
| Boots | | | | | |

Key: Each shoe = 1 vote

You survey your friends about their favorite kinds of pets. The choices are cat, dog, bird, or fish. Draw a symbol for each choice. (Encuestas a tus amigos acerca de sus tipos de mascotas favoritas. Las opciones son gatos, perros, aves o peces. Dibuja un símbolo para cada opción.)

| | | | |
|---|---|---|---|
| | | | |
| | | | |
| | | | |
| | | | |

What picture can be used to show data in the picture graph? (¿Qué imagen puede usarse para mostrar datos en la gráfica con imágenes?)

**Favorite Subject**

| | | | | | |
|---|---|---|---|---|---|
| Math | | | | | |
| Science | | | | | |
| Reading | | | | | |
| Art | | | | | |

Key: Each book = 1 vote

Dinah Zike's
**Visual
Kinesthetic
Vocabulary**®

Chapter 10

✂ cut on all dashed lines    ▭ fold on all solid lines

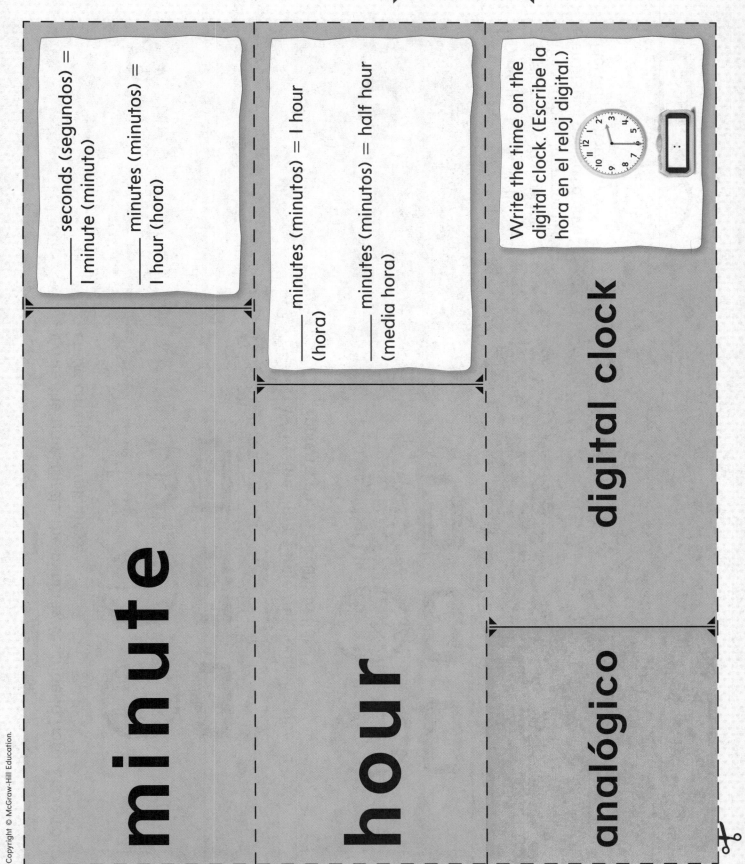

_____ seconds (segundos) =
1 minute (minuto)

_____ minutes (minutos) =
1 hour (hora)

_____ minutes (minutos) = 1 hour
(hora)

_____ minutes (minutos) = half hour
(media hora)

Write the time on the
digital clock. (Escribe la
hora en el reloj digital.)

**minute**

**hour**

**digital clock**

**analógico**

ra

o

Draw the hands on the analog clock to show the time. (Dibuja las manecillas en el reloj analógico para mostrar la hora.)

9:30

reloj digital

analog

Write the time. Circle the minutes. (Escribe la hora. Encierra en un círculo los minutos.)

Write the time. Circle the hour. (Escribe la hora. Encierra en un círculo la hora.)

Dinah Zike's
**VKV**
Visual
Kinesthetic
Vocabulary®

Chapter II

✂ cut on all dashed lines          ⬜ fold on all solid lines

When you estimate, you... (Cuando estimas, tú...)

_____ centimeters (centímetros) = I meter (metro)

Which would you measure using centimeters? (¿Cuál medirías usando centímetros?)

length of racetrack (la longitud de una pista de carreras)

width of a book (el ancho de un libro)

height of a house (la altura de una casa)

estimate

centimeter

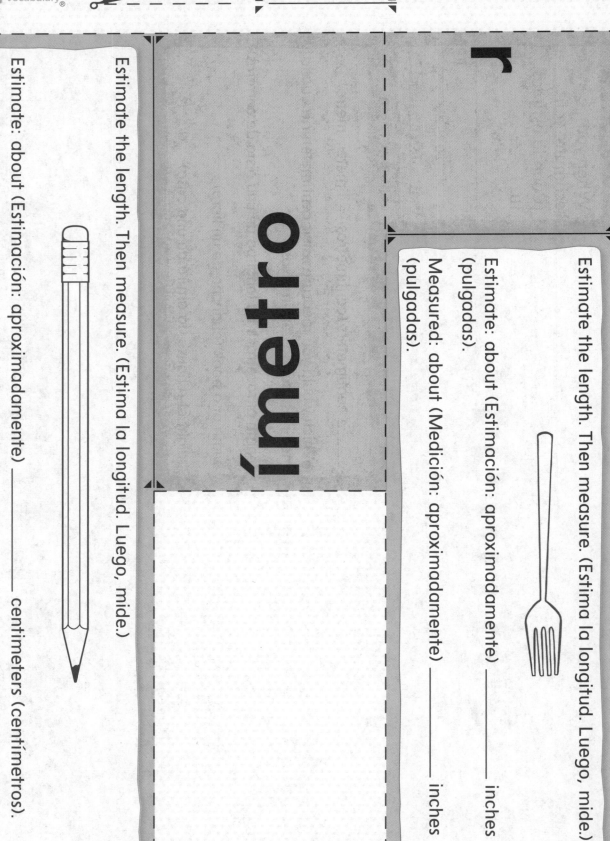

Estimate the length. Then measure. (Estima la longitud. Luego, mide.)

Estimate: about (Estimación: aproximadamente) ———— centimeters (centímetros).

Measured: about (Medición: aproximadamente) ———— centimeters (centímetros).

ímetro

r

Estimate the length. Then measure. (Estima la longitud. Luego, mide.)

Estimate: about (Estimación: aproximadamente) ———— inches (pulgadas).

Measured: about (Medición: aproximadamente) ———— inches (pulgadas).

Dinah Zike's
**Visual
Kinesthetic
Vocabulary**®

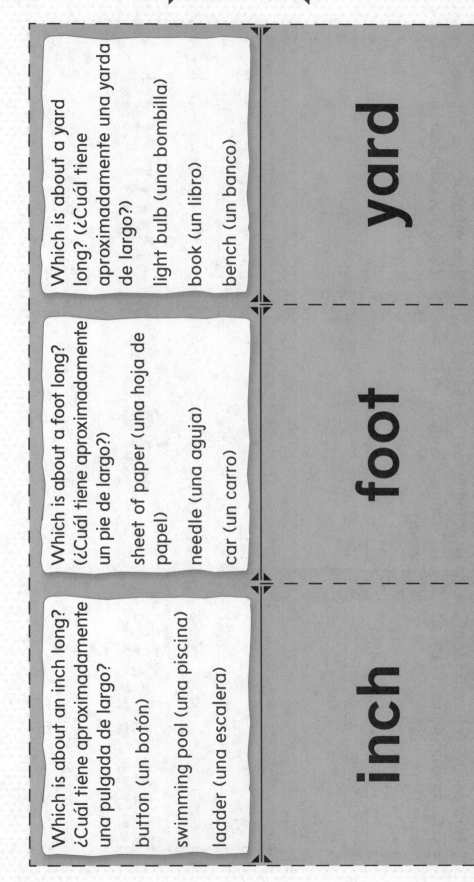

Which is about a yard long? (¿Cuál tiene aproximadamente una yarda de largo?)

light bulb (una bombilla)

book (un libro)

bench (un banco)

**yard**

Which is about a foot long? (¿Cuál tiene aproximadamente un pie de largo?)

sheet of paper (una hoja de papel)

needle (una aguja)

car (un carro)

**foot**

Which is about an inch long? ¿Cuál tiene aproximadamente una pulgada de largo?

button (un botón)

swimming pool (una piscina)

ladder (una escalera)

**inch**

Dinah Zike's
**V K V** Visual
Kinesthetic
Vocabulary®

Chapter II

✂ cut on all dashed lines          ⬚ fold on all solid lines

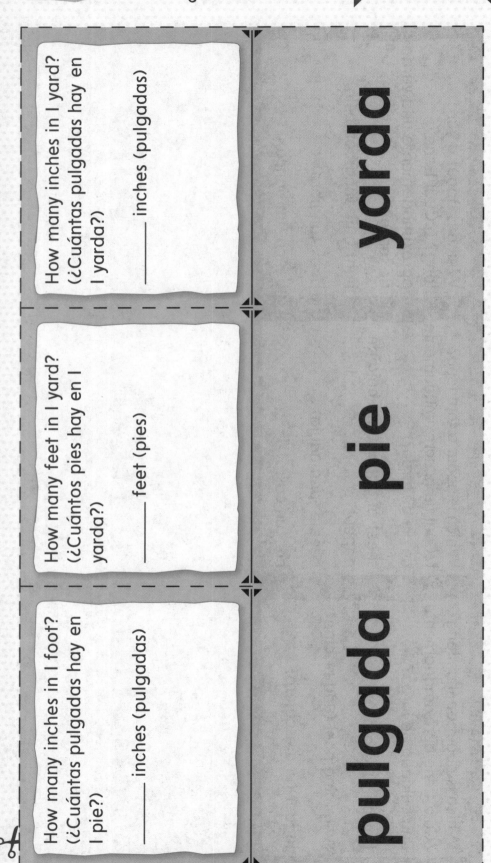

How many inches in 1 yard?
(¿Cuántas pulgadas hay en 1 yarda?)

_____ inches (pulgadas)

How many feet in 1 yard?
(¿Cuántos pies hay en 1 yarda?)

_____ feet (pies)

How many inches in 1 foot?
(¿Cuántas pulgadas hay en 1 pie?)

_____ inches (pulgadas)

yarda

pie

pulgada

Dinah Zike's
**VKV** Visual
Kinesthetic
Vocabulary®

Chapter 12

✂ cut on all dashed lines          fold on all solid lines

How is a trapezoid similar to a rectangle? (¿En qué se parece un trapecio a un rectángulo?)

_____

How is it different? (¿En qué se diferencia?)

_____

A pentagon has _____ sides and _____ angles.

(Un pentágono tiene _____ lados y _____ ángulos.)

**trapezoid**

**hexá**

**pentagon**

Trace the shape. (Dibuja la figura.)

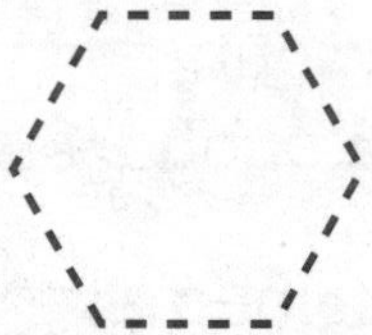

Trace the shape.
(Dibuja la figura.)

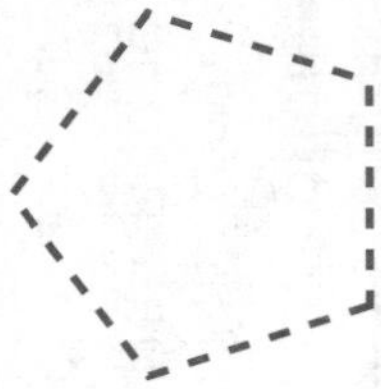

cio

pentágono

Circle the trapezoid. (Encierra en un círculo el trapecio.)

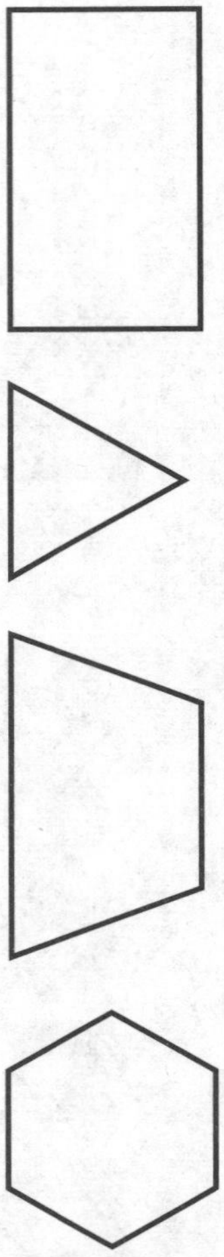

hexa

A hexagon has
——— sides and
——— angles.

(Un hexágono tiene
——— lados y
——— ángulos.)

Dinah Zike's
**VKV** Visual
Kinesthetic
Vocabulary ®

Chapter 12

✂ cut on all dashed lines          ▭ fold on all solid lines

A cube is a (two-dimensional, three-dimensional) shape. Each face of a cube is a (rectangle, square). (Un cubo es una figura (bidimensional, tridimensional). Cada cara de un cubo es un (rectángulo, cuadrado).)

A cone has —— face and —— vertex.

(Un cono tiene —— cara y —— vértice.)

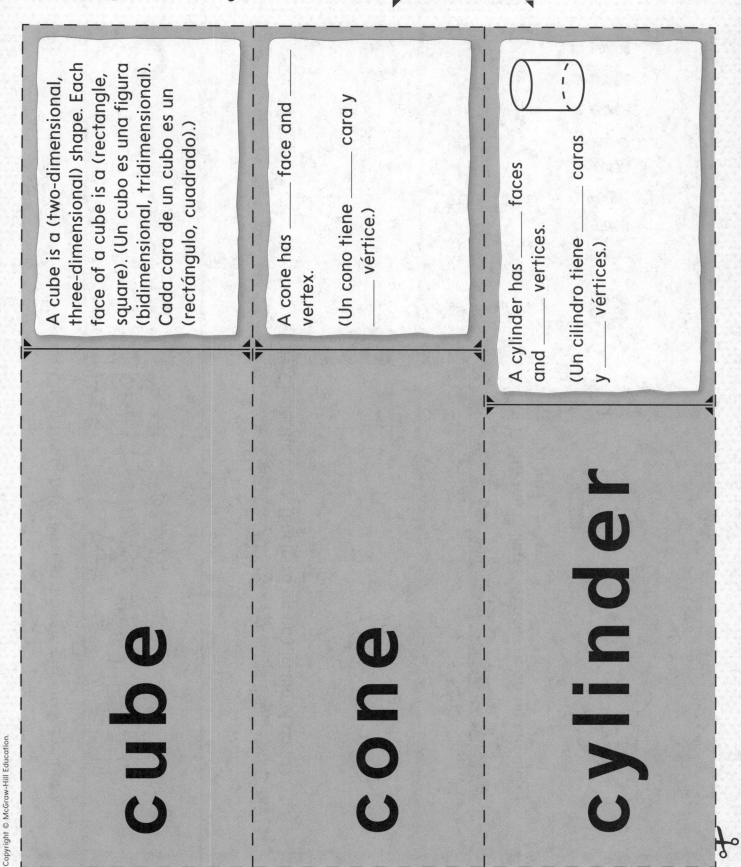

A cylinder has —— faces and —— vertices.

(Un cilindro tiene —— caras y —— vértices.)

# cube

# cone

# cylinder

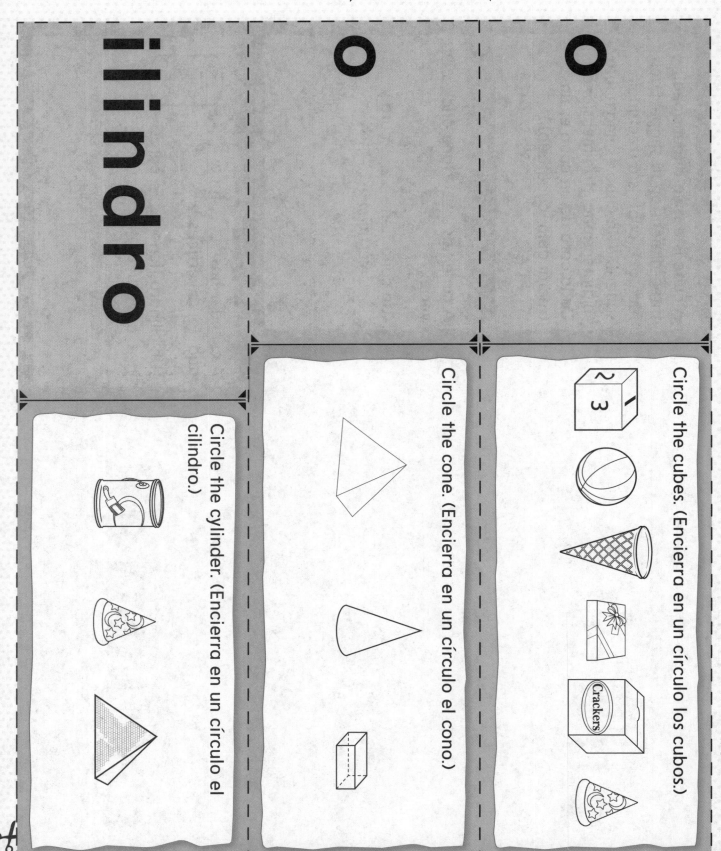

ilindro

O

O

Circle the cylinder. (Encierra en un círculo el cilindro.)

Circle the cone. (Encierra en un círculo el cono.)

Circle the cubes. (Encierra en un círculo los cubos.)

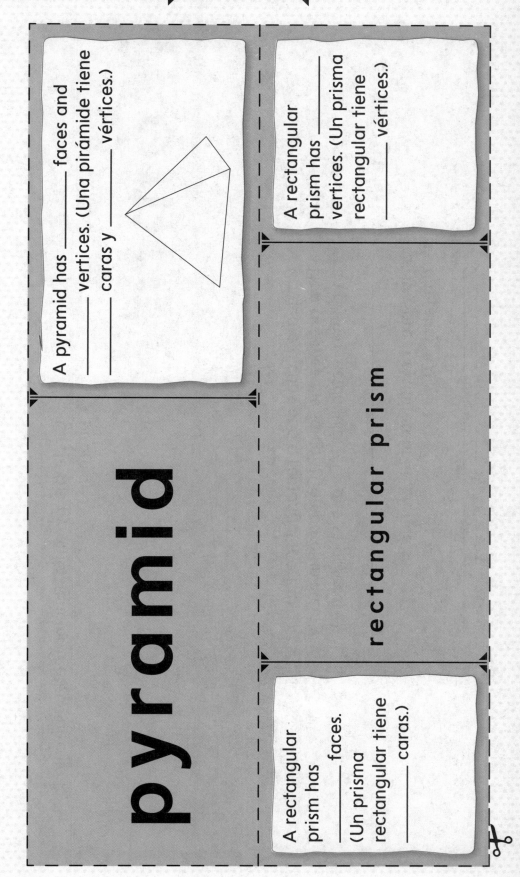

A pyramid has _____ faces and _____ vertices. (Una pirámide tiene _____ caras y _____ vértices.)

A rectangular prism has _____ vertices. (Un prisma rectangular tiene _____ vértices.)

**pyramid**

**rectangular prism**

A rectangular prism has _____ faces. (Un prisma rectangular tiene _____ caras.)

Dinah Zike's
# VKV
Visual
Kinesthetic
Vocabulary ®

Chapter 12

✂ cut on all dashed lines

📄 fold on all solid lines

rectangular

# irámide

Compare a cube and a rectangular prism.
How are they the same? (Compara un cubo y
un prisma rectangular. ¿En qué se parecen?)

How are they different? (¿En qué se
diferencian?)

Circle the pyramid. (Encierra en un círculo la
pirámide.)

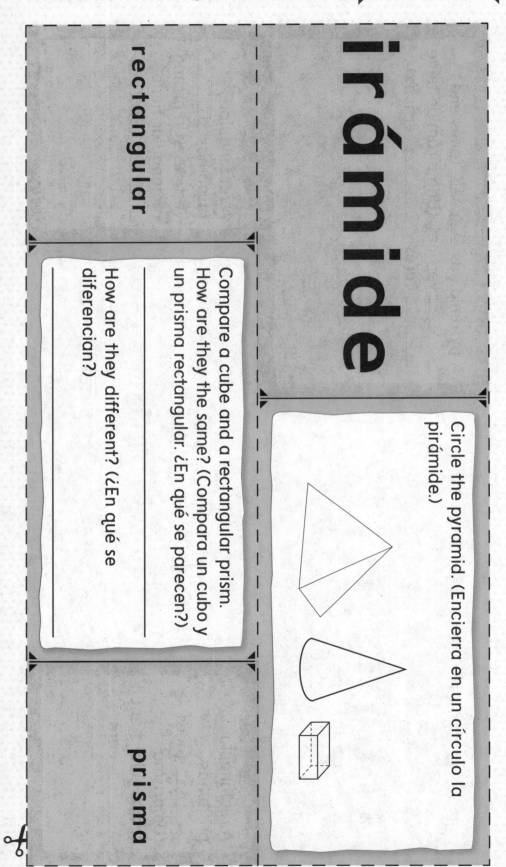

prisma

# VKV Answer Appendix

## Chapter 1
### VKV3
*sum:* 1 + 3 = 4 [anno circle around 4]
*add/subtract:* 5 + 3 = 8; 4 + 1 = 5;
2 + 7 = 9, 8 − 2 = 6; 5 − 0 = 5;
8 − 4 = 4

### VKV4
*suma:* 7, 4, 9, 8
*sumar/restar:* 7, 8, 9, 7; 3, 8, 4, 6

### VKV5
*sum/difference:* addition; subtraction
*doubles:* 12, 8, 16

### VKV6
*suma/diferencia:* 9, 13, 11, 12; 10, 7, 6, 6
*dobles:* 9 beads; near doubles

### VKV7
*count on:* 9, 12, 10, 10
*count back:* 6, 7, 2, 3

### VKV8
*seguir contando:* 11 apples
*contra hacia atrás:* 8 birds

## Chapter 2/3
### VKV9
*odd/even number:* See students' work.
*regroup:* there are 10 r more ones.

### VKV10
*reagrupar:* 15 + 8

## Chapter 5
### VKV11
*digit:* 0, 1, 2, 3, 4, 5, 6, 7, 8, 9
*place value:* 379

### VKV12
*dígito:* See students' work.
*valor posicional:* 500 + 20 + 3 = 523;
700 + 0 + 6 = 706

### VKV13
*hundreds/thousands:* 1,527, 844, 25,
1,013, 649, 79; 527, 1,415, 2,365, 13, 48,
942
*compare:* greatest

### VKV14
*centenas/millares:* 416, 132, 785; 1,412,
1,123
*comparar:* <, >, >, =, <, >

### VKV15
*greater than/equal to/less than:* See
students' work.

### VKV16
*mayor que/igual a/menor que:* >, <,
>, <, =, <, <, >, >, >

## Chapter 8
### VKV17
### VKV18
*signo de dólar:* <, >, >, =
*moneda de 1¢:* 1¢ + 1¢ + 1¢ = 3¢

### VKV19
### VKV20
*moneda de 5¢:* 5¢ + 5¢ + 5¢ + 5¢ =
20¢
*moneda de 10¢:* 10¢ + 10¢ = 20¢
*moneda de 25¢:* 25¢ + 25¢ + 25¢ =
75¢

# Chapter 9

## VKV21
*picture graph:* 4, 3, 2
*symbol:* Sample answers: beach, bicycling, rollerblading, baseball
*data:* numbers or symbols that show information.

## VKV22
*gráfica con imágenes:* See students' work.
*símbolo:* See students' work.
*datos:* The number of students who choose each subject as their favorite.

# Chapter 10

## VKV23
*minute:* 60; 60
*hour:* 60; 30
*digital clock:* 2:30

## VKV24
*minuto:* 4:30, 7:00, 10:30
*hora:* 12:00, 4:30, 1:30
*reloj digital:* See students' work.

# Chapter 11

## VKV25
*estimate:* find a number close to an exact amount.
*centimeter:* 100; width of a book

## VKV26
*estimar:* See students' work.
*centímetro:* See students' work.

## VKV27
*inch/foot/yard:* button; sheet of paper; bench

## VKV28
*pugada/pie/yarda:* 12 inches; 3 feet; 36 inches

# Chapter 12

## VKV29
*trapezoid:* Both have 4 sides and 4 angles. A trapezoid has only one pair of equal sides.
*pentagon:* 5 sides, 5 angles; See students' work.

## VKV30
*trapecio:* See students' work.
*pentágono:* 6 sides, 6 angles; See students' work.

## VKV31
*cube:* three-dimensional; square
*cone:* 1 face, 1 vertex
*cylinder:* 2 faces, 0 vertices

## VKV32
*cubo:* See students' work.
*cono:* See students' work.
*cilindro:* See students' work.

## VKV33
*pyramid:* 5 faces, 5 vertices
*rectangular prism:* 6 faces; 8 vertices

## VKV34
*pirámide:* See students' work.
*prisma rectangular:* Both have 6 faces and 8 vertices. The faces of a cube are all squares. The faces of a rectangular prism are rectangles.